Ballistic Resistance of Body Armor

BALLISTIC RESISTANCE
OF BODY ARMOR

UNITED STATES DEPARTMENT OF JUSTICE

Nova Science Publishers, Inc.
New York

Copyright © 2009 by Nova Science Publishers, Inc.

All rights reserved. No part of this book may be reproduced, stored in a retrieval system or transmitted in any form or by any means: electronic, electrostatic, magnetic, tape, mechanical photocopying, recording or otherwise without the written permission of the Publisher.

For permission to use material from this book please contact us:
Telephone 631-231-7269; Fax 631-231-8175
Web Site: http://www.novapublishers.com

NOTICE TO THE READER

The Publisher has taken reasonable care in the preparation of this book, but makes no expressed or implied warranty of any kind and assumes no responsibility for any errors or omissions. No liability is assumed for incidental or consequential damages in connection with or arising out of information contained in this book. The Publisher shall not be liable for any special, consequential, or exemplary damages resulting, in whole or in part, from the readers' use of, or reliance upon, this material.

Independent verification should be sought for any data, advice or recommendations contained in this book. In addition, no responsibility is assumed by the publisher for any injury and/or damage to persons or property arising from any methods, products, instructions, ideas or otherwise contained in this publication.

This publication is designed to provide accurate and authoritative information with regard to the subject matter covered herein. It is sold with the clear understanding that the Publisher is not engaged in rendering legal or any other professional services. If legal or any other expert assistance is required, the services of a competent person should be sought. FROM A DECLARATION OF PARTICIPANTS JOINTLY ADOPTED BY A COMMITTEE OF THE AMERICAN BAR ASSOCIATION AND A COMMITTEE OF PUBLISHERS.

LIBRARY OF CONGRESS CATALOGING-IN-PUBLICATION DATA

Available Upon Request
ISBN 978-1-60692-295-8

Published by Nova Science Publishers, Inc. ✢New York

FOREWORD

This document, NIJ Standard–0101.06, *"Ballistic Resistance of Body Armor,"* is a minimum performance standard developed in collaboration with the Office of Law Enforcement Standards (OLES) of the National Institute of Standards and Technology (NIST). It is produced as part of the Standards and Testing Program of the National Institute of Justice (NIJ), Office of Justice Programs, U.S. Department of Justice. This standard is a technical document that specifies the minimum performance requirements that equipment must meet to satisfy the requirements of criminal justice agencies and the methods that shall be used to test this performance. This standard is used by the NIJ Voluntary Compliance Testing Program (CTP) to determine which body armor models meet the minimum performance requirements for inclusion on the NIJ Compliant Products List. In addition, manufacturers, criminal justice agencies, and others may use the tests described in this standard to determine whether a particular armor design meets their own requirements. Users are strongly encouraged to have this testing conducted in accordance with the NIJ CTP. Procurement officials may also refer to this standard in their purchasing documents and require that equipment offered for purchase meet or exceed these requirements.

This document is a testing and performance standard and provides precise and detailed test methods. Additional requirements, processes, and procedures for CTP participants are detailed in the NIJ Voluntary CTP Administrative Procedures Manual. Those seeking guidance concerning the selection and application of body armor should refer to the most recent revision of the *Selection and Application Guide to Personal Body Armor, NIJ Guide 100*, which is published as a separate document and explains in nontechnical language how to select equipment that provides the level of performance required by a purchasing agency.

Publication of this revision of the standard does not invalidate or render unsuitable any body armor models previously determined by NIJ to be compliant to either the NIJ 2005 Interim Requirements or the NIJ Standard–0101.04 Rev. A Requirements. While it may not necessary to remove these existing armors from service, agencies are advised to always require their procurements to meet or exceed the most recent and up-to-date version of this standard.

Personal body armor that is independently tested to this standard by manufacturers, purchasers, or other parties will not be considered as NIJ compliant unless the body armor is submitted and tested through the NIJ CTP and found in compliance with this standard.

When manufacturers seek NIJ compliance of their armor to this standard and the armor contains unique materials or forms of construction that may not have been anticipated when this standard was drafted, NIJ may modify the test methods of the standard to take those features into account. If NIJ determines that the model meets the requirements of the standard, modified test methods, and the NIJ CTP, NIJ will include the armor on the NIJ Compliant Products List.

NIJ standards are subjected to continued research, development, testing, review, and revision. This standard and its successors will be re-evaluated and modified as necessary. Because of ongoing advancements in ballistic-resistant materials, changes to armor designs and manufacturing techniques, and improvements in testing methods, NIJ reserves the right to withhold NIJ compliance status for body armor that meets the current requirements but is deemed by NIJ to pose a risk to officer safety. NIJ also reserves the right to alter or modify existing test methods and/or requirements to address perceived weaknesses in varying designs of body armor submitted for inclusion on the NIJ Compliant Products List.

Technical comments and recommended revisions are welcome. Please send all written comments and suggestions to the Deputy Director, Office of Science and Technology, National Institute of Justice, Office of Justice Programs, U.S. Department of Justice, 810 Seventh Street, NW, Washington, DC 20531. Before citing this or any other NIJ standard in a contract document, users should verify that the most recent edition of the standard is used. The most recent edition will be available on the Justice Technology Information Network (JUSTNET) at http://www.justnet.org, or write to the Director, Office of Law Enforcement Standards, National Institute of Standards and Technology, 100 Bureau Drive, Stop 8102, Gaithersburg, MD 20899.

John Morgan, Deputy Director for Science and
Technology, National Institute of Justice

CONTENTS

Commonly Used Symbols And Abbreviations	xi
1. Purpose And Scope	1
2. NIJ Body Armor Classification	3
2.1 Type IIA (9 mm; .40 S&W)	3
2.2 Type II (9 mm; .357 Magnum)	3
2.3 Type IIIA (.357 SIG; .44 Magnum)	3
2.4 Type III (Rifles)	4
2.5 Type IV (Armor Piercing Rifle)	4
2.6 Special Type	4
3. Definitions	5
4. Sample Requirements and Laboratory Configuration	11
4.1 Test Samples	11
4.1.1 Flexible Vests and Jackets	11
4.1.2 Hard Armors and Plate Inserts	12
4.1.3 Accessory Ballistic Panels	17
4.1.4 Workmanship	17
4.1.5 Labeling	17
4.1.6 Armor Carriers With Removable Ballistic Panels	20
4.1.7 Armors With Built-In Inserts or Trauma Packs	20
4.2 Laboratory Configuration and Test Equipment	20
4.2.1 Range Configuration	20
4.2.2 Test Rounds and Barrels	22
4.2.3 Velocity Measurement Equipment	22
4.2.4 Armor Submersion Equipment	23
4.2.5 Armor Backing Material	23
5. Flexible Armor Conditioning Protocol	27
5.1 Purpose and Scope	27
5.2 Pretest Parameters	27
5.2.1 Storage of Armors	27
5.2.2 Pretest Calibrations	27
5.2.3 Test Conditions	27
5.3 Laboratory Configuration and Test Equipment	28
5.3.1 General Parameters	28
5.3.2 Controls	29
5.3.3 Test Interruption	29
5.3.4 Procedure	30
6. Hard Armor Conditioning Protocol	33
6.1 Purpose and Scope	33
6.2 Pretest Parameters	33
6.2.1 Storage of Armors	33
6.2.2 Pretest Calibrations	33
6.2.3 Test Conditions	33
6.3 Laboratory Configuration and Test Equipment	35

		6.3.1	General Parameters	35
		6.3.2	Controls	35
		6.3.3	Test Interruption	35
		6.3.4	Conditioning Procedure	36

7. Ballistic Test Methods 39
  7.1 Purpose and Scope 39
  7.2 Test Order 39
  7.3 Workmanship Examination 39
    7.3.1 Armor Carriers and Ballistic Panel Covers 39
    7.3.2 Ballistic Panels 39
    7.3.3 Label Examination 39
  7.4 Sampling 40
  7.5 Sample Acclimation 40
    7.5.1 Inserts 40
  7.6 Fair Hit Requirements for All Ballistic Tests 40
    7.6.1 Minimum Shot-to-Edge Distance 40
    7.6.2 Minimum Shot-to-Shot Distance 40
  7.7 Backing Material Preparation and Sample Mounting for All Ballistic Tests 41
    7.7.1 Backing Material Fixture Preparation 41
    7.7.2 Mounting Armor for Ballistic Testing 41
  7.8 Perforation and Backface Signature Test (P-BFS) 43
    7.8.1 Shot Location Marking 43
    7.8.2 Armor Submersion 45
    7.8.3 Test Threats for P-BFS Tests 45
    7.8.4 Test Duration 45
    7.8.5 Requirements for Number of Shots and Number of Armor Samples 45
    7.8.6 P-BFS Test for Special Type Armor 49
    7.8.7 P-BFS Test for Accessory Ballistic Panels 49
    7.8.8 P-BFS Performance Requirements 49
  7.9 Ballistic Limit (BL) Determination Test 50
    7.9.1 Requirements for Number of BL Tests and Test Samples 51
    7.9.2 Test Procedure Requirements 51
    7.9.3 Backing Material Conditioning 52
    7.9.4 Data Set Tabulation 52
    7.9.5 Ballistic Limit Performance Requirements 52
8. References 53
APPENDIX A – Acceptable Bullets for Handloading 55
APPENDIX B – Common Special Type Threats 57
APPENDIX C – Armor Sizing Templates 59
APPENDIX D – Analysis of Backface Signature Data 65
APPENDIX E – Analysis of Ballistic Limit Data 69
APPENDIX F – Explanatory Materials 71
Index 75

PREFACE

The Standards and Testing Program is sponsored by the Office of Science and Technology of the National Institute of Justice (NIJ), Office of Justice Programs, U.S. Department of Justice. The program responds to the mandate of the Homeland Security Act of 2002, which directed the Office of Science and Technology to establish and maintain performance standards in accordance with the National Technology Transfer and Advancement Act of 1995 (Public Law 104–113) to test and evaluate law enforcement technologies that may be used by Federal, State, and local law enforcement agencies. The Homeland Security Act of 2002 also directed the Office of Science and Technology to establish and maintain a program to certify, validate, and mark or otherwise recognize law enforcement technology products that conform to the standards mentioned above.

The Standards and Testing Program is a basic and applied research effort that determines the technological needs of justice system agencies, sets minimum performance standards for specific devices, tests commercially available equipment against those standards, and disseminates the standards and the test results to criminal justice agencies nationally and internationally.

The *Office of Law Enforcement Standards* (OLES) at the National Institute of Standards and Technology develops voluntary national performance standards for compliance testing to ensure that individual items of equipment are suitable for use by criminal justice agencies. The standards are based upon laboratory testing and evaluation of representative samples of each item of equipment to determine the key attributes, develop test methods, and establish minimum performance requirements for each essential attribute. In addition to the technical standards, OLES also produces technical reports and user guidelines that explain in nontechnical terms the capabilities of available equipment.

The *National Law Enforcement and Corrections Technology Center* (NLECTC), operated by a grantee, coordinates a national compliance testing program conducted by independent laboratories. The standards developed by OLES serve as performance benchmarks against which commercial equipment is measured.

ABBREVIATIONS

STANDARD SPECIFIC ABBREVIATIONS

ACP =	Automatic Colt Pistol	LR =	Long Rifle
ANSI =	American National Standards Institute	LRN =	Lead Round Nose
AP =	Armor Piercing	NLECTC =	National Law Enforcement and Corrections Technology Center
BFS =	Backface Signature		
BL =	Ballistic Limit	P-BFS =	Perforation and Backface Signature
CPO =	Compliance Program Office	RN =	Round Nose
CTP =	Compliance Testing Program	S&W =	Smith & Wesson
CTR =	Compliance Test Report	SAAMI =	Sporting Arms and Ammunition Manufacturers' Institute
FMJ =	Full Metal Jacket		
ISO =	International Standards Organization		
JHP =	Jacketed Hollow Point	SJHP =	Semi Jacketed Hollow Point
JSP =	Jacketed Soft Point	SJSP =	Semi Jacketed Soft Point

COMMONLY USED SYMBOLS AND ABBREVIATIONS

A	ampere	H	henry	nm	nanometer
ac	alternating current	h	hour	No.	number
AM	amplitude modulation	hf	high frequency	o.d.	outside diameter
cd	candela	Hz	hertz	Ω	ohm
cm	centimeter	i.d.	inside diameter	p.	page
CP	chemically pure	in	inch	Pa	pascal
c/s	cycle per second	IR	infrared	pe	probable error
d	day	J	joule	pp.	pages
dB	decibel	L	lambert	ppm	parts per million
dc	direct current	L	liter	qt	quart
°C	degree Celsius	Lb	pound	rad	radian
°F	degree Fahrenheit	lbf	pound force	rf	radio frequency
diam	diameter	lbf·in	pound force inch	rh	relative humidity
emf	electromotive force	lm	lumen	s	second
eq	equation	ln	logarithm (base e)	SD	standard deviation
F	farad	log	logarithm (base 10)	sec.	section
fc	footcandle	M	molar	SWR	standing wave ratio
fig.	figure	m	meter	uhf	ultrahigh frequency
FM	frequency modulation	min.	minute	UV	ultraviolet
ft	foot	mm	millimeter	V	volt
ft/s	foot per second	mph	miles per hour	vhf	very high frequency
g	acceleration	m/s	meter per second	W	watt
g	gram	N	newton	λ	wavelength
gr	grain	N·m	newton meter	wt	weight

area = unit^2 (e.g., ft^2, in^2, etc.); volume = unit^3 (e.g., ft^3, m^3, etc.)

PREFIXES

d	deci (10^{-1})		da	deka (10)
c	centi (10^{-2})		h	hecto (10^2)
m	milli (10^{-3})		k	kilo (10^3)
μ	micro (10^{-6})		M	mega (10^6)
n	nano (10^{-9})		G	giga (10^9)
p	pico (10^{-12})		T	tera (10^{12})

COMMON CONVERSIONS
(See ASTM E380)

0.30480 m = 1 ft   4.448222 N = 1 lbf
2.54 cm = 1 in   1.355818 J = 1 ft·lbf
0.4535924 kg = 1 lb   0.1129848 N.m = 1 lbf·in
0.06479891 g = 1 gr   14.59390 N/m = 1 lbf/ft
0.9463529 L = 1 qt   6894.757 Pa = 1 lbf/in^2
3600000 J = 1 kW·h   1.609344 km/h = 1 mph

Temperature: $T_{°C} = (T_{°F} - 32) \times 5/9$
Temperature: $T_{°F} = (T_{°C} \times 9/5) + 32$

Purpose And Scope

The purpose of this standard is to establish minimum performance requirements and test methods for the ballistic resistance of personal body armor intended to protect against gunfire.

This standard is a revision of NIJ Standard–0101.04, dated September 2000. It supersedes the NIJ 2005 Interim Requirements, dated September 2005, NIJ Standard–0101.04, and all other revisions and addenda to NIJ Standard–0101.04.

The scope of the standard is limited to ballistic resistance only; this standard does not address threats from knives and sharply pointed instruments, which are different types of threats and are addressed in the current version of NIJ Standard–0115 *Stab Resistance of Personal Body Armor*.

Body armor manufacturers and purchasers may use this standard to help to determine whether specific armor models meet the minimum performance standards and test methods identified in this document. However, NIJ strongly encourages body armor manufacturers to participate in the NIJ Voluntary Compliance Testing Program (CTP) and encourages purchasers to insist that the armor model(s) they purchase be tested by the NIJ CTP and be listed on the NIJ Compliant Products List. This will help to assure that the armor models will meet the minimum performance standards for use by the criminal justice community.

The ballistic tests described in this standard have inherent hazards. Adequate safeguards for personnel and property must be employed when conducting these tests.

NIJ BODY ARMOR CLASSIFICATION

Personal body armor covered by this standard is classified into five types (IIA, II, IIIA, III, IV) by level of ballistic performance. In addition, a special test class is defined to allow armor to be validated against threats that may not be covered by the five standard classes.

The classification of an armor panel that provides two or more levels of NIJ ballistic protection at different locations on the ballistic panel shall be that of the minimum ballistic protection provided at any location on the panel.

2.1 Type IIA (9 mm; .40 S&W)

Type IIA armor that is new and unworn shall be tested with 9 mm Full Metal Jacketed Round Nose (FMJ RN) bullets with a specified mass of 8.0 g (124 gr) and a velocity of 373 m/s ± 9.1 m/s (1225 ft/s ± 30 ft/s) and with .40 S&W Full Metal Jacketed (FMJ) bullets with a specified mass of 11.7 g (180 gr) and a velocity of 352 m/s ± 9.1 m/s (1155 ft/s ± 30 ft/s).

Type IIA armor that has been conditioned shall be tested with 9 mm FMJ RN bullets with a specified mass of 8.0 g (124 gr) and a velocity of 355 m/s ± 9.1 m/s (1165 ft/s ± 30 ft/s) and with .40 S&W FMJ bullets with a specified mass of 11.7 g (180 gr) and a velocity of 325 m/s ± 9.1 m/s (1065 ft/s ± 30 ft/s).

2.2 Type II (9 mm; .357 Magnum)

Type II armor that is new and unworn shall be tested with 9 mm FMJ RN bullets with a specified mass of 8.0 g (124 gr) and a velocity of 398 m/s ± 9.1 m/s (1305 ft/s ± 30 ft/s) and with .357 Magnum Jacketed Soft Point (JSP) bullets with a specified mass of 10.2 g (158 gr) and a velocity of 436 m/s ± 9.1 m/s (1430 ft/s ± 30 ft/s).

Type II armor that has been conditioned shall be tested with 9 mm FMJ RN bullets with a specified mass of 8.0 g (124 gr) and a velocity of 379 m/s ±9.1 m/s (1245 ft/s ± 30 ft/s) and with .357 Magnum JSP bullets with a specified mass of 10.2 g (158 gr) and a velocity of 408 m/s ±9.1 m/s (1340 ft/s ± 30 ft/s).

2.3 Type IIIA (.357 SIG; .44 Magnum)

Type IIIA armor that is new and unworn shall be tested with .357 SIG FMJ Flat Nose (FN) bullets with a specified mass of 8.1 g (125 gr) and a velocity of 448 m/s ± 9.1 m/s (1470 ft/s ± 30 ft/s) and with .44 Magnum Semi Jacketed Hollow Point (SJHP) bullets with a specified mass of 15.6 g (240 gr) and a velocity of 436 m/s ± 9.1 m/s (1430 ft/s ± 30 ft/s).

Type IIIA armor that has been conditioned shall be tested with .357 SIG FMJ FN bullets with a specified mass of 8.1 g (125 gr) and a velocity of 430 m/s ± 9.1 m/s (1410 ft/s ± 30 ft/s) and with .44 Magnum SJHP bullets with a specified mass of 15.6 g (240 gr) and a velocity of 408 m/s ± 9.1 m/s (1340 ft/s ± 30 ft/s).

2.4 Type III (Rifles)

Type III hard armor or plate inserts shall be tested in a conditioned state with 7.62 mm FMJ, steel jacketed bullets (U.S. Military designation M80) with a specified mass of 9.6 g (147 gr) and a velocity of 847 m/s ± 9.1 m/s (2780 ft/s ± 30 ft/s).

Type III flexible armor shall be tested in both the "as new" state and the conditioned state with 7.62 mm FMJ, steel jacketed bullets (U.S. Military designation M80) with a specified mass of 9.6 g (147 gr) and a velocity of 847 m/s ± 9.1 m/s (2780 ft/s ± 30 ft/s).

For a Type III hard armor or plate insert that will be tested as an *in conjunction* design, the flexible armor shall be tested in accordance with this standard and found compliant as a stand-alone armor at its specified threat level. The combination of the flexible armor and hard armor/plate shall then be tested as a system and found to provide protection at the system's specified threat level. NIJ-approved hard armors and plate inserts must be clearly labeled as providing ballistic protection only when worn in conjunction with the NIJ-approved flexible armor system with which they were tested.

2.5 Type IV (Armor Piercing Rifle)

Type IV hard armor or plate inserts shall be tested in a conditioned state with .30 caliber armor piercing (AP) bullets (U.S. Military designation M2 AP) with a specified mass of 10.8 g (166 gr) and a velocity of 878 m/s ± 9.1 m/s (2880 ft/s ± 30 ft/s).

Type IV flexible armor shall be tested in both the "as new" state and the conditioned state with .30 caliber AP bullets (U.S. Military designation M2 AP) with a specified mass of 10.8 g (166 gr) and a velocity of 878 m/s ± 9.1 m/s (2880 ft/s ± 30 ft/s).

For a Type IV hard armor or plate insert that will be tested as an *in conjunction* design, the flexible armor shall be tested in accordance with this standard and found compliant as a stand-alone armor at its specified threat level. The combination of the flexible armor and hard armor/plate shall then be tested as a system and found to provide protection at the system's specified threat level. NIJ-approved hard armors and plate inserts must be clearly labeled as providing ballistic protection only when worn in conjunction with the NIJ-approved flexible armor system with which they were tested.

2.6 Special Type

A purchaser having a special requirement for a level of protection other than one of the above standard types and threat levels should specify the exact test round(s) and reference measurement velocities to be used and indicate that this standard shall govern all other aspects. Guidance on common special type threats and the appropriate threat velocities is provided in appendix B, along with a methodology for determining the correct reference velocity for other threats.

DEFINITIONS

3.1 Absolute Humidity: The quantity of water in a given volume of air, often reported in grams per cubic meter.

3.2 Accessory Ballistic Panels: Accessory panels are armor components that are detachable or removable from the primary body armor sample and are intended to provide comparable ballistic protection. Examples of accessory panels include groin, coccyx, and side protection panels, which are attached to or inserted into the external armor carrier but are not integral to the armor panels or armor sample.

3.3 Angle of Incidence: The angle between the bullet's line of flight and the perpendicular to the front surface of the backing material fixture as shown in figure 1.

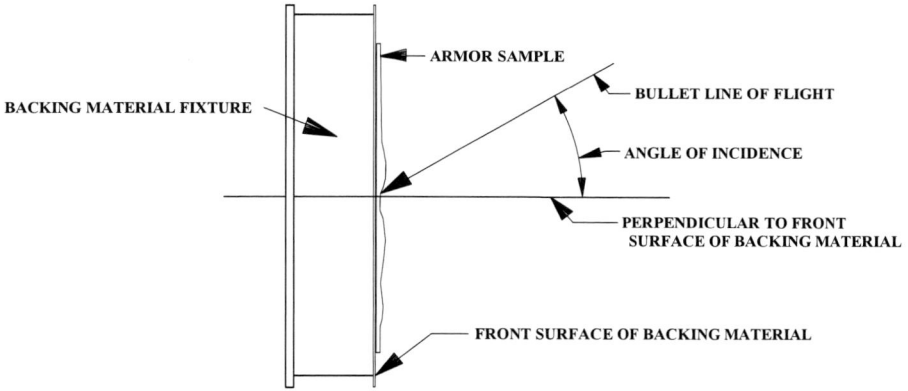

Figure 1. Angle of incidence

3.4 Armor Carrier: A component of the armor sample or armor panel whose primary purpose is to retain the ballistic panel and provide a means of supporting and securing the armor garment to the user. Generally, the carrier is not ballistic resistant.

3.5 Armor Conditioning: Environmental and mechanical conditioning of armor prior to ballistic testing, which consists of exposure to specified conditions of temperature, humidity, and mechanical damage.

3.6 Armor Panel or Panel: The portion of an armor sample that consists of an external ballistic cover and its internal ballistic panel. The word "panel," if not preceded by the word "ballistic," refers to an armor panel in this standard.

3.7 Armor Sample: One complete armor garment. Typically, a front armor panel, a back armor panel, and the armor carrier comprise a body armor sample. The armor sample may be a single wraparound style or consist of multiple parts that are worn around the body.

3.8 Backface Signature (BFS): The greatest extent of indentation in the backing material caused by a nonperforating impact on the armor. The BFS is the perpendicular distance between two planes, both of which are parallel to the front surface of the backing material fixture. One plane contains the reference point on the original (pretest) backing material surface that is co-linear with the bullet line of flight. (If armor were not present, the bullet

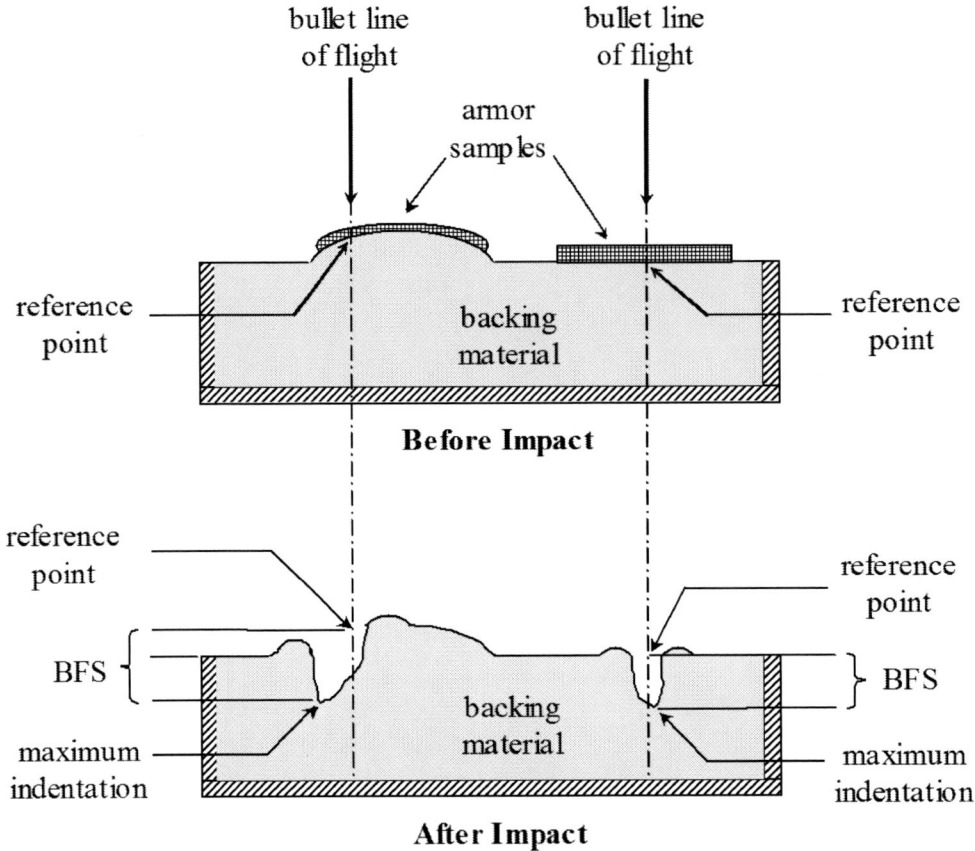

Figure 2. Examples of BFS measurements

would strike this point.) The other plane contains the point that represents the deepest indentation in the backing material. Depending on bullet–armor–backing material interactions, the two points that define the locations of the measurement planes may not be co-linear with the bullet line of flight. Examples of how BFS is measured are shown in figure 2.

3.9 Backing Material: A homogenous block of nonhardening, oil-based modeling clay placed in contact with the back of the armor panel during ballistic testing.

3.10 Backing Material Fixture: A box-like rigid frame, with a removable back, which contains the backing material. The removable back is used for perforation-backface signature testing but is not used for ballistic limit testing.

3.11 Ballistic Limit: For a given bullet type, the velocity at which the bullet is expected to perforate the armor 50 % of the time. The ballistic limit is typically denoted as the V50 or V_{50} value.

3.12 Ballistic Panel: The protective component of an armor sample or armor panel, consisting primarily of ballistic resistant materials. The ballistic panel is usually enclosed in a non-removable panel cover, which is considered part of the ballistic panel. The ballistic panel is normally enclosed within an armor carrier.

3.13 Baseline Ballistic Limit: The experimentally derived ballistic limit of an armor model when new.

3.14 Body Armor: An item of personal protective equipment that provides protection against specific ballistic threats within its coverage area. In this standard, the term body armor refers to that which provides coverage primarily for the torso.

3.15 Complete Penetration: This nomenclature is no longer used in this standard. See "perforation." *Perforation* replaces *complete penetration*. Although the terms *partial penetration* and *complete penetration* are no longer used in this standard, they may still be used by test laboratories for compatibility with military standards.

3.16 Compliance Test Group: A group of armor samples submitted for testing according to this standard.

3.17 Condensation: Precipitation of water vapor on a surface whose temperature is lower than the dew point of the ambient air. The dew point depends on the quantity of water vapor in the air. The dew point, the absolute humidity, and the vapor pressure are directly interdependent. Condensation occurs on a test item when the temperature at the surface of the item placed in the test chamber is lower than the dew point of the air in the chamber. As a result, the item may need to be preheated to prevent condensation.

3.18 Dewpoint (or dew point): The temperature to which a parcel of air must be cooled (at constant barometric pressure) for water vapor present in the air to condense into water (dew).

3.19 Fair Hit: The impact of a bullet on an armor panel that meets the shot spacing and velocity requirements of section 7.6.

3.20 Flexible Body Armor: Body armor constructed of pliable, textile-based materials such that the complete system is capable of being flexed. Such systems are typically in the form of vests or jackets that provide greater coverage area than rigid plate armor. Generally, these armors provide protection against handgun threats.

3.21 Full Metal Jacketed Bullet (FMJ): A bullet consisting of a lead core completely covered, except for the base, with copper alloy (approximately 90 % copper and 10 % zinc). "Total Metal Jacket (TMJ)," "Totally Enclosed Metal Case (TEMC)," and other commercial terminology for bullets with electro-deposited copper and copper alloy coatings have been tested and are considered comparable to FMJ bullets for this standard.

3.22 Hard Armor or Rigid Armor: Rigid armor systems, plates, inserts, accessories, or semi-rigid armor systems constructed with rigid plates that are typically designed to provide protection against rifle threats.

3.23 **In Conjunction Armor**: A combination of either two flexible armor panels or a flexible armor with a plate insert that is designed to provide increased stab or ballistic protection.

For an *in conjunction* armor system designed with a separate flexible stab armor panel insert added to the flexible ballistic panel and intended to provide dual threat protection against stab and ballistic threats, the complete system must be tested in the end-use configuration to meet the requirements of this standard. If the flexible ballistic armor is designed to be worn separately from the stab armor panel, the flexible armor shall be tested in accordance with this standard and found compliant as a stand-alone armor at its specified threat level.

For an *in conjunction* armor designed to meet the requirements of NIJ Standard–0101.06, the flexible armor shall be tested in accordance with this standard and found compliant as a stand-alone armor at its specified threat level. The combination of the flexible armor and hard armor/plate shall then be tested as a system and found to provide protection at the system's specified threat level. NIJ-approved hard armors and plate inserts must be clearly labeled as providing ballistic protection only when worn in conjunction with the NIJ-approved flexible armor system with which they were tested.

3.24 **In Conjunction Plate**: An insert that is designed to provide increased ballistic or stab protection only when it is used <u>with a particular model</u> of a flexible armor vest or jacket.

3.25 **Insert**: A removable or nonremovable unit of ballistic material that can enhance the ballistic performance of the armor panel in a localized area but not over the entire area intended for ballistic protection. Some inserts are known as *trauma packs*, *trauma plates*, or *trauma inserts*, but other forms of inserts are possible.

3.26 **Jacketed Hollow Point Bullet (JHP)**: A bullet consisting of a lead core that has a hollow cavity or hole located in the nose of the bullet and is completely covered, except for the hollow point, with a copper alloy (approximately 90 % copper and 10 % zinc) jacket.

3.27 **Jacketed Soft Point Bullet (JSP)**: A lead bullet that is completely covered, except for the nose, with a copper alloy (approximately 90 % copper and 10 % zinc) jacket. This bullet is also known as a Semi Jacketed Soft Point (SJSP).

3.28 **Lead Bullet**: A bullet made entirely of lead, which may be alloyed with hardening agents.

3.29 **Maximum Velocity**: The specified reference velocity for a given threat round (table 4) plus 9.1 m/s (30 ft/s).

3.30 **Minimum Velocity**: The specified reference velocity for a given threat round (table 4) minus 9.1 m/s (30 ft/s).

3.31 **Nonplanar Armor**: Body armor with features that prevent the armor from making full contact with the backing material surface. Examples include female body armor with bust cups and curved rigid plates.

Definitions

3.32 Panel: See *Armor Panel*.

3.33 Penetration: Any impact where the projectile passes into the armor is a penetration. A penetration may either be a *stop* or a *perforation*. The result is considered a *stop* or *partial penetration* if (1) the entire projectile is captured or deflected by the armor, and (2) no fragments of either the projectile or the armor become embedded in the backing material. If any part of the projectile passes through the armor, the result is considered a *perforation* or a *complete penetration*. Although the terms *partial penetration* and *complete penetration* are no longer used in this standard, they may still be used by test laboratories for compatibility with military standards.

3.34 Perforation: Any impact that creates a hole passing through the armor. This may be evidenced by any of the following: (1) the presence of the projectile, a projectile fragment, or an armor fragment in the clay backing material; (2) a hole that passes through the armor and/or backing material; or (3) any portion of the bullet being visible from the body side of the armor panel. The term *perforation* is synonymous with the term *complete penetration*.

3.35 Plate Inserts: Hard armor plates or semi-rigid plates that are intended to be inserted into pockets of flexible vests and jackets to provide increased protection, particularly to provide protection against rifle threats.

3.36 Reference Velocity: The specified measurement velocity goal for test rounds used in perforation-backface signature ballistic performance tests.

3.37 Relative Humidity: The ratio of the amount of water in a given parcel of air at a given temperature to the maximum amount of water that the air can hold at that temperature.

3.38 Rigid Armor or Systems: See 3.22 *Hard Armor*.

3.39 Round Nose Bullet (RN): A bullet with a blunt or rounded nose. A bullet with a generally blunt or rounded nose or tip, which possesses a small flat surface at the tip of the bullet, shall also be considered a round nose bullet for this standard.

3.40 Semi Jacketed Hollow Point Bullet (SJHP): A bullet consisting of a lead core with a copper alloy (approximately 90 % copper and 10 % zinc) jacket covering the base and bore riding surface (major diameter), which leaves some portion of the lead core exposed, thus forming a lead nose or tip, which has a hollow cavity or hole located in the nose or tip of the bullet.

3.41 Semi Jacketed Soft Point Bullet (SJSP): A bullet, also known as a Jacketed Soft Point (JSP), consisting of a lead core with a copper alloy (approximately 90 % copper and 10 % zinc) jacket covering the base and bore riding surface (major diameter), which leaves some portion of the lead core exposed, thus forming a lead nose or tip.

3.42 Shot-to-Edge Distance: The distance from the center of the bullet impact to the nearest edge of the ballistic panel.

3.43 Shot-to-Shot Distance: The distance from the center of the bullet impact to the center of the nearest prior bullet impact.

3.44 Stop: The outcome of a shot where the projectile is either captured or deflected by the armor, with no portion of the projectile or fragments of the armor *perforating* the armor.

3.45 Striking Device: A device used to establish an appropriate, flat reference surface for the backing material (see 3.9 *Backing Material*).

3.46 Strike Face: The surface of an armor sample or panel designated by the manufacturer as the surface that should face the incoming ballistic threat. Also, the side of the armor opposite the wear face (see 3.51 *Wear Face*).

3.47 Test Series: The set of all shots necessary to obtain the required number of fair hits on a single armor panel or plate, as defined in table 4, and the set of all shots necessary to generate ballistic limit response data, as defined in table 8.

3.48 Textile-Based Materials: Materials manufactured by weaving or felting yarns into a fabric, or by embedding or laminating fibers in sheets of plastic film.

3.49 Trauma Insert/Pack/Plate: See 3.25 *Insert*.

3.50 Vapor Pressure: The pressure exerted by a vapor in equilibrium with its solid or liquid phase.

3.51 Wear Face: The side of the armor that is worn against the body. Also, the side of the armor opposite the strike face (see 3.46 *Strike Face*).

3.52 Yaw: The angular deviation of the longitudinal axis of the projectile from its line of flight, measured as close to the target as practical.

SAMPLE REQUIREMENTS AND LABORATORY CONFIGURATION

For an armor model to comply with this standard, all requirements of this section shall be met, and the appropriate number of samples shall be subjected to the tests defined in section 5, section 6, and section 7.

4.1 Test Samples

4.1.1 Flexible Vests and Jackets

For flexible armor in the form of concealable or tactical vests or jackets, 14 complete armors per test threat, sized as described in section 4.1.1.1, shall constitute the compliance test group.

Types IIA through IIIA armor shall be tested with two test threats and the compliance test group will consist of 28 complete armors. Types III and IV armor shall be tested with a single test threat, and the compliance test group will consist of 14 complete armors. If special testing is to be performed with additional test threats, the compliance test group will require an additional 14 complete armors for each additional test threat.

Refer to table 1, figure 3, figure 4, and the following sections for details on the required sample sizes and the tests to be performed on each. Additional samples may be required if optional tests are to be performed or if all necessary testing cannot be completed on the standard compliance test group.

4.1.1.1 Test Sample Sizes

Two sizes are required in the compliance test group: larger sized and smaller sized. The sizes of the armor samples will depend on the range of the sizes over which the armor model will be produced. Sample armor sizing templates are provided in appendix C. Table 1 summarizes the quantities required for each armor template size.

Larger sized test samples: The armor manufacturer shall determine which template size most closely matches the largest size of the armor model that the manufacturer will produce.

Table 1. Armor samples required for each test threat (flexible vests and jackets)

(Note that quantities below should be doubled for types with 2 threats [Types IIA – IIIA])

Armor Template Size	Number Required	Armor Condition		Ballistic Tests		
		New	Conditioned	P-BFS	Ballistic Limit	Spare
Larger	11	8	New	2	5	1
		3	Conditioned	1	1	1
Smaller	3	2	New	2	-	-
		1	Conditioned	1	-	-
Total	14	10	4	6	6	2

For each test threat, the manufacturer shall provide 11 armor samples that fit this selected template. These armors will be referred to as the larger sized test samples.

Smaller sized test samples: The armor manufacturer shall determine which template size most closely matches the smallest size of the armor model that the manufacturer will produce. For each test threat, the manufacturer shall provide three armor samples that fit this selected template. These armors will be referred to as the smaller sized test samples.

4.1.1.2 Sample Utilization

Conditioned Samples: Refer to table 1, figure 3, and figure 4. From each group of 14 armors, one small sized armor and three large sized armors shall be selected at random and subjected to the armor conditioning protocol described in section 5. After the selected samples have been conditioned, the one small sized sample and one randomly selected large sized sample shall be subjected to the Perforation and Backface Signature (P-BFS) test described in section 7. Of the remaining two large sized samples, one shall be randomly selected and subjected to the Ballistic Limit (BL) test described in section 7. The remaining sample is a spare.

New Samples: Refer to table 1, figure 3, and figure 4. From each group of 14 armors, two small sized armors and two large sized armors shall be selected at random and subjected to the P-BFS test described in section 7. Five large sized armors will be selected at random and subjected to the BL test described in section 7. The remaining sample is a spare.

4.1.2 Hard Armors and Plate Inserts

All hard armors and plate inserts shall be subjected to a 24 shot P-BFS test and to either a 24 shot or a 12 shot BL test. All hard armors and plate inserts shall be conditioned per section 6 prior to ballistic testing. All hard armors and plate inserts shall be no larger than 254 mm x 305 mm (10.0 in x 12.0 in) for testing. The required number of armor samples is dependent on the armor type, as described in the following sections.

4.1.2.1 Type III

For hard armors and insert plates intended to provide Type III protection, the compliance test group shall consist of nine armor panels. The armor panels shall be sufficiently large to allow for a minimum of six shots per panel. These requirements are outlined in figure 5. Four armor panels will be used for the P-BFS testing described in section 7. A minimum of four armor panels will be subjected to the BL test described in section 7, with a minimum of 24 shots. The remaining armor panel is a spare and be used if necessary.

4.1.2.2 Type IV

For hard armors and insert plates intended to provide Type IV protection, the compliance test group shall consist of a sufficient number of armor panels to allow a 24 shot P-BFS test and a 12 shot BL test, with at least one spare armor panel. These requirements are outlined in figure 6. For armor models capable of withstanding only a single ballistic impact, the compliance test group shall consist of 37 armor panels. For armor models capable of withstanding multiple ballistic impacts, the manufacture shall specify the number of shots to be fired at each panel, and the compliance test group size may be reduced accordingly. However, a maximum of six P-BFS shots may be taken on a single panel. The compliance test group must include one armor panel as a spare.

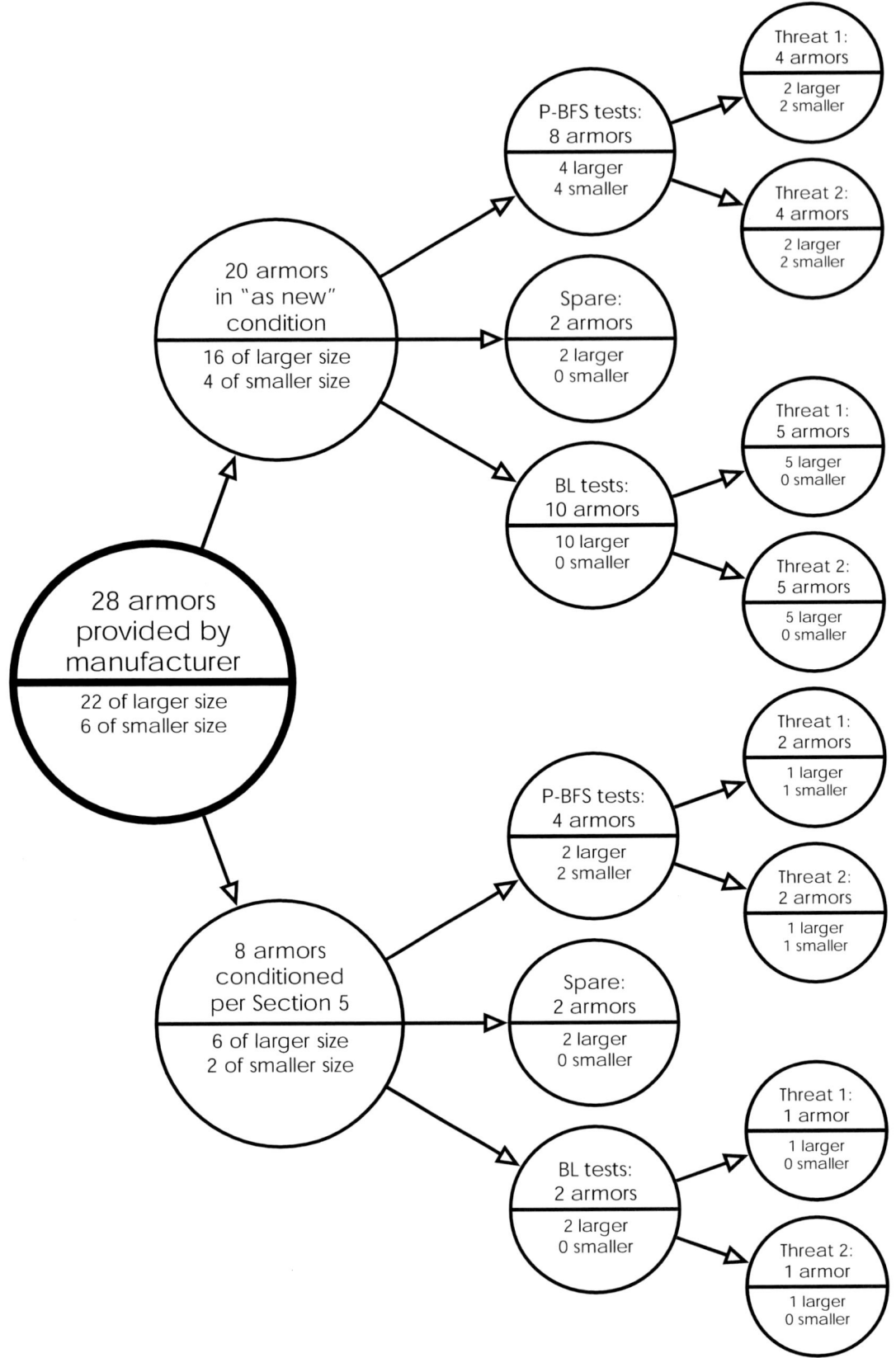

Figure 3. Sample quantity and utilization for armor Types IIA, II, and IIIA

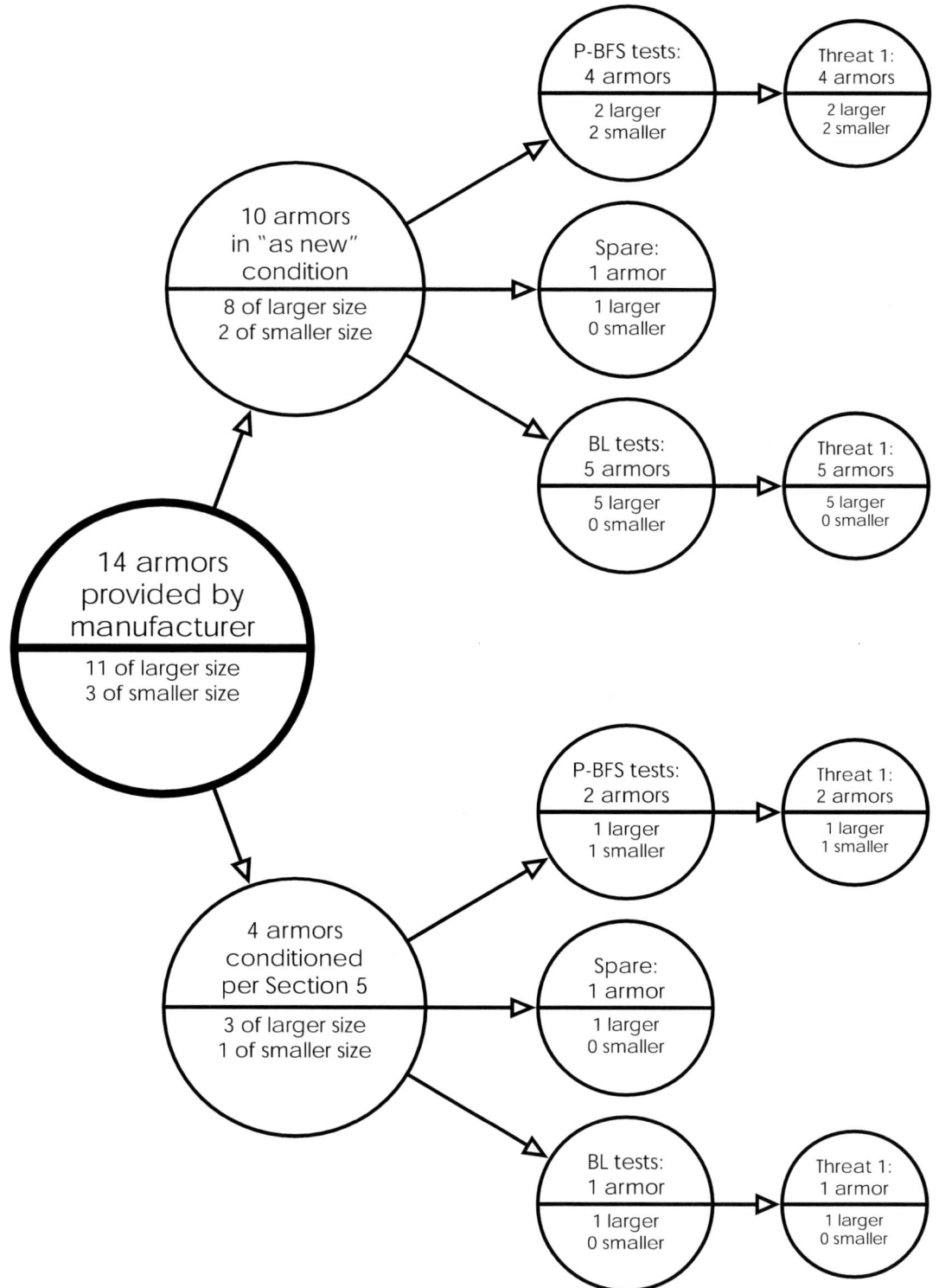

Figure 4. Sample quantity and utilization for flexible armor of Type III, Type IV, and Type Special

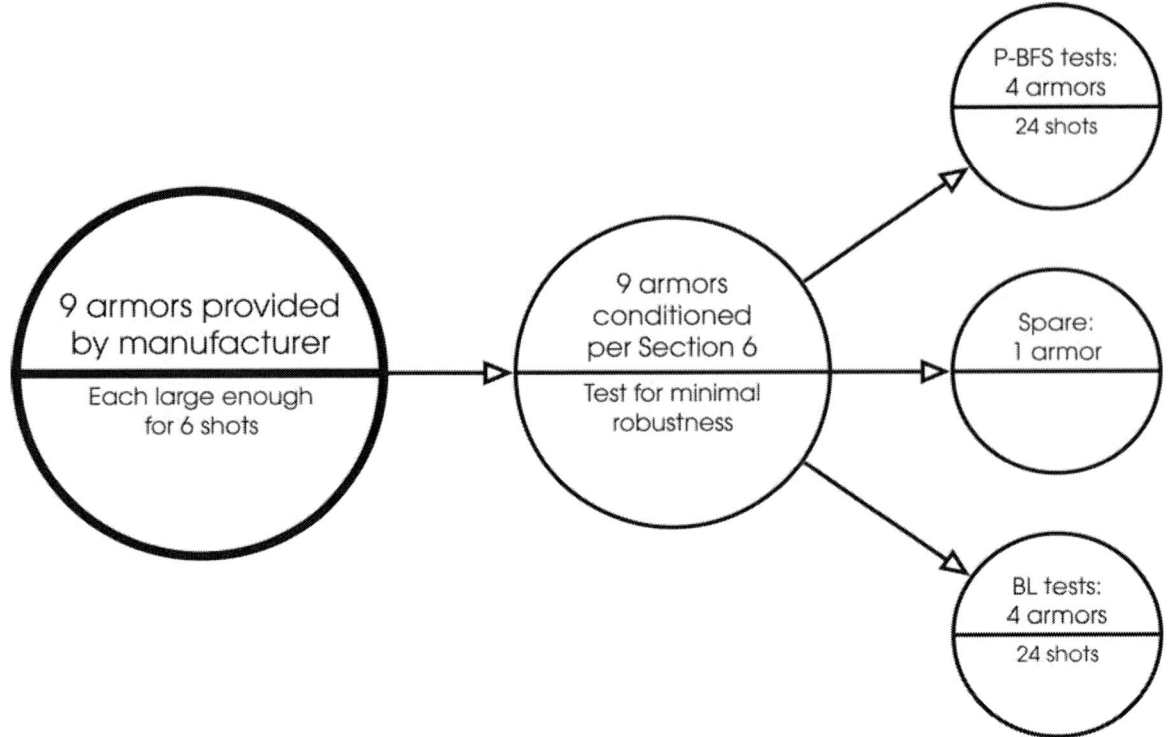

Figure 5. Sample quantity and utilization for hard armors and plate inserts of Type III

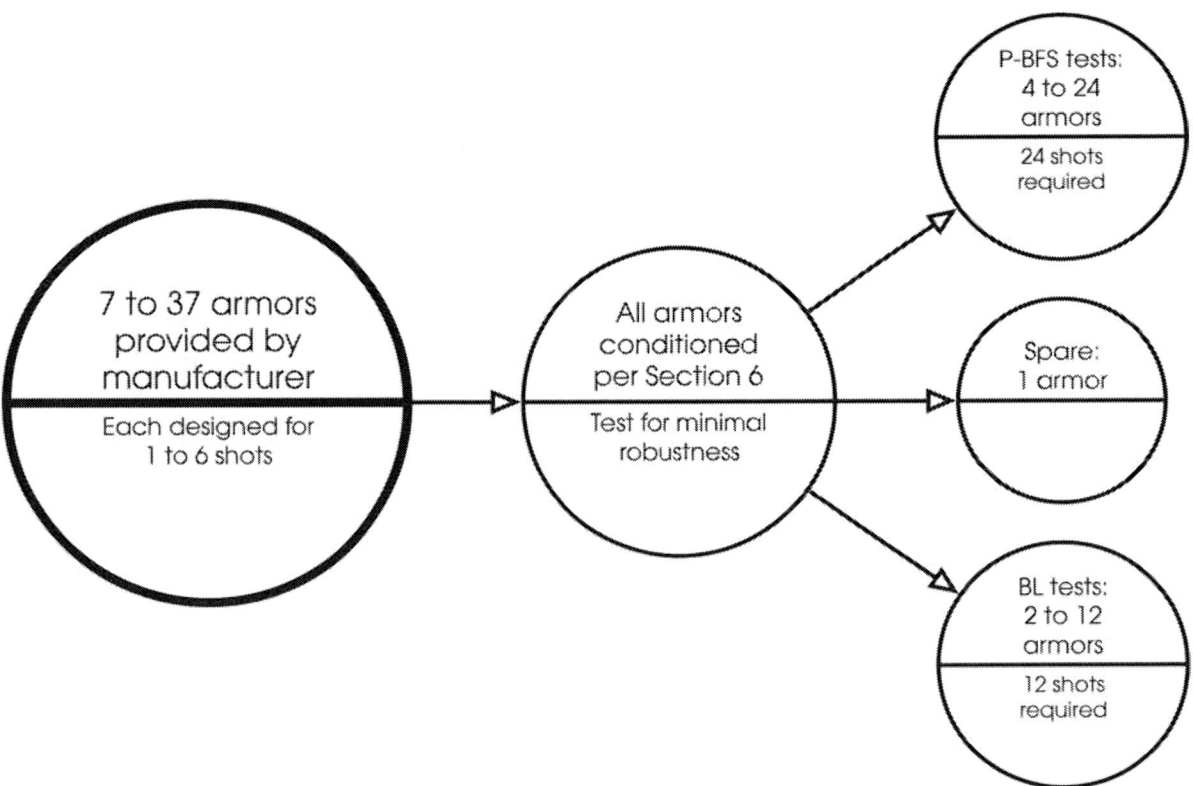

Figure 6. Sample quantity and utilization for hard armors and plate inserts of Type IV

4.1.3 Accessory Ballistic Panels

All accessory ballistic panels shall be subjected to a 24 shot P-BFS test for each threat round. For groin and coccyx protectors, eight samples are required for each threat round, and each sample will be subjected to at least three shots. For other types of accessory panels, the required number of armor samples will be dependent on the panel size, and the number of impacts per test sample shall be determined by the fair hit impact spacing criteria for flexible armor panels. A maximum of six P-BFS shots may be taken on a single panel.

4.1.4 Workmanship

All armor shall be free from evidence of inferior workmanship, such as wrinkles, blisters, cracks or fabric tears, fraying, crazing, or chipped or sharp corners and edges. Stitching must be straight and secure. All samples shall be identical in appearance and manner of construction. All samples of flexible vests and jackets shall be sized to meet the appropriate sizing templates, and all armors in each size group shall be identically sized. There shall be no variance in construction details between individual armor samples or between any armor sample and the manufacturer's documentation.

4.1.5 Labeling

Armor shall be durably and clearly marked (labeled) in a readable type and font size, with a pretest label in accordance with the requirements set forth below. An example is shown in figure 7.

4.1.5.1 Ballistic Panels

Every ballistic panel cover shall have a label. The label shall be permanently attached to the face of the panel. The label shall contain the following information (figure 7):

(a) Name, logo or other identification of the manufacturer.
(b) The rated level of protection, according to section 2 of this standard and referenced to this edition of the standard (i.e., Type II in accordance with NIJ Standard–0101.06).
(c) A test ID number or model designation that uniquely identifies the armor design for testing purposes.
(d) The panel size.
(e) A sample number or serial number that uniquely identifies each armor sample.
(f) Location of manufacture.
(g) The date of manufacture.
(h) Identification of the proper orientation of the ballistic panel, i.e., strike face or wear face.
(i) A warning in 14 point or larger type and a minimum of 1 ½ times larger than the rest of the type on the label, exclusive of the information required in (a) above, stating that the armor panel is a test sample that has not yet been demonstrated to provide ballistic resistance and that the armor panel is not intended to be worn. Printing color changes are acceptable but cannot be substituted for the type size requirement herein.

These labeling requirements apply to armor samples of models that are being submitted for precompliance or compliance testing.

4.1.5.1.1 Labeling Requirements for Armors Listed on the NIJ Compliant Products List

Armor samples of models that have been listed on the NIJ Compliant Products List and that are being retested, lot tested, etc. shall be labeled in accordance with the requirements of the Compliance Testing Program (CTP). Label requirements for armors listed on the NIJ Compliant Products List are provided in the NIJ Voluntary CTP Administrative Procedures Manual.

In addition to the requirements of the CTP, all armor panels of listed models shall include the following:

(a) The rated level of protection, according to section 2 of this standard and referenced to this edition of the standard (i.e., Type II in accordance with NIJ Standard–0101.06).
(b) Strike face or wear face. The proper orientation of the ballistic panel in the carrier must be clearly identified.
(c) For Types IIA, II, and IIIA armor, a warning in type not less than 14 point and a minimum of 1 ½ times larger than the rest of the type on the label, exclusive of the manufacturer identification and logo, stating that the armor is not intended to protect the wearer from rifle fire.
(d) If applicable, a warning in type not less than 14 point and a minimum of 1 ½ times larger than the rest of the type on the label, exclusive of the manufacturer identification and logo, stating that the armor is not intended to protect the wearer from sharp edged or pointed instruments. This warning may be combined with the warning in (c) above.

4.1.5.2   Armor Carriers With Nonremovable Ballistic Panels

Armor with ballistic panels that are nonremovable shall, in addition to the label required for the ballistic panel, have a label on the carrier (figure 7) that is in conformance with the requirements for the ballistic panels, unless the armor is so constructed that the ballistic panel label is not covered by the carrier. The label may be positioned in a location where it is not visible when the armor is worn, but it shall be easy to locate and easily readable when the armor is donned and doffed.

4.1.5.3   Label Permanency and Durability

All labels shall be sufficiently permanent and durable so that they will remain securely attached to the armor or carrier and be fully legible for the life of the armor. The durability of the label markings shall be checked with the following procedure:

[1] A representative area of the label markings shall be rubbed by hand for 15 s with a cotton cloth soaked with distilled water.
[2] The same area shall then be rubbed by hand for 15 s with a cotton cloth soaked with denatured alcohol (methylated spirit).
[3] Finally, the same area shall then be rubbed by hand for 15 s with cotton cloth soaked with isopropyl alcohol.

The label markings within the representative area must remain legible.

This test shall be performed on either a stand-alone example label or on a label of a spare armor sample. **This test shall not be performed on an armor sample that will subsequently undergo ballistic testing, unless approved by the armor manufacturer.**

The labels of armor samples subjected to the armor conditioning protocol in section 5 must remain legible after the conditioning has been completed; however, these samples will not be subjected to the durability test described in the preceding paragraphs.

Company Name
Company Address
[Company Logo - Optional]

Warning: Test Article

This armor panel is intended for testing. *It is **not** intended for personal use. It has **not** been demonstrated to have any ballistic resistance.*

Do Not Wear!

Test ID: _____
Size: _____
Location of Manufacture: _____
Date of Manufacture: _____/_____/_____
Lot Number: _____
Sample Number: _____

Threat Level (NIJ 0101.06): _____

Warning: Test Article

This armor panel is intended for testing. *It is **not** intended for personal use. It has **not** been demonstrated to have any ballistic resistance.*

Do Not Wear!

** Wear Face **
This side to be positioned against backing material.

Figure 7. Example label for test samples

4.1.6 Armor Carriers With Removable Ballistic Panels

Armor with ballistic panels that are removable shall be submitted for testing with cotton or polycotton carriers with an areal density of not more than 250 g/m^2 (7.37 oz/yd^2). The test carriers shall not have strapping, strapping attachment points, pockets for accessory plates or trauma packs, or any accessory mounting points, with the following exceptions:

(a) Armors submitted for testing in conjunction with Type III or Type IV plates shall have the necessary pockets for the plates with which they will be tested.
(b) Armors that are to be tested using their own strapping for mounting and support during the tests shall have appropriate strapping and strapping attachment points. If the strapping or strapping attachment materials are deemed by the CTP to have a significant influence on the ballistic performance, then the armor shall be treated as if the ballistic panels were nonremovable and should be tested in carriers of the same construction as will be used for production armors.
(c) Armor carriers for samples that will undergo the conditioning protocol described in section 5 shall not have strapping, strapping attachment points, or any accessory mounting points. For armors that are to be tested using their own strapping for mounting and support, separate carriers may be provided for the conditioning protocol and ballistic testing.

4.1.7 Armors With Built-In Inserts or Trauma Packs

For armor models that contain built-in inserts or trauma packs, manufacturers must submit a detailed diagram of the location of each trauma pack. During the P-BFS test, the shot locations shall be adjusted so that areas other than the built-in inserts or trauma packs are tested.

4.2 Laboratory Configuration and Test Equipment

The armor shall be tested in a facility and with test equipment that meets the following requirements.

4.2.1 Range Configuration

4.2.1.1 Ambient Test Conditions

The ambient conditions shall be recorded before and after each armor panel firing sequence (6 or 12 shots) and, unless otherwise specified, shall be as stated below.

(a) Temperature: 21 °C ± 2.9 °C (70 °F ± 5 °F).
(b) Relative humidity: 50 % ± 20 %.

4.2.1.2 Range Preparation

The test equipment will be arranged as shown in figure 8. For handgun rounds, the armor panel shall be mounted 5.0 m ± 1.0 m (16.4 ft ± 3.28 ft) from the muzzle of the test barrel, and for rifle rounds, the armor panel should be mounted 15 m ± 1.0 m (49.2 ft ± 3.28 ft) from the muzzle of the test barrel, with the following exception. In order to minimize the possibility of excessive yaw at impact, or for other range configuration reasons, the distance may be adjusted for each threat; however, the distance shall not be less than 4 m (13.1 ft) for any round. In the case of rifle rounds, if the distance is adjusted to less than 14 m (45.9 ft), the bullet yaw shall be experimentally verified to confirm that the angle of incidence is within 5° of the intended angle.

The backing material fixture should be rigidly held by a suitable test stand, which shall permit the entire armor and backing material assembly to be shifted vertically and horizontally such that the entire face of the backing material can be targeted.

4.2.1.3 Instrumentation

Before testing, all electronic equipment will be allowed sufficient time to warm up so that stability is achieved.

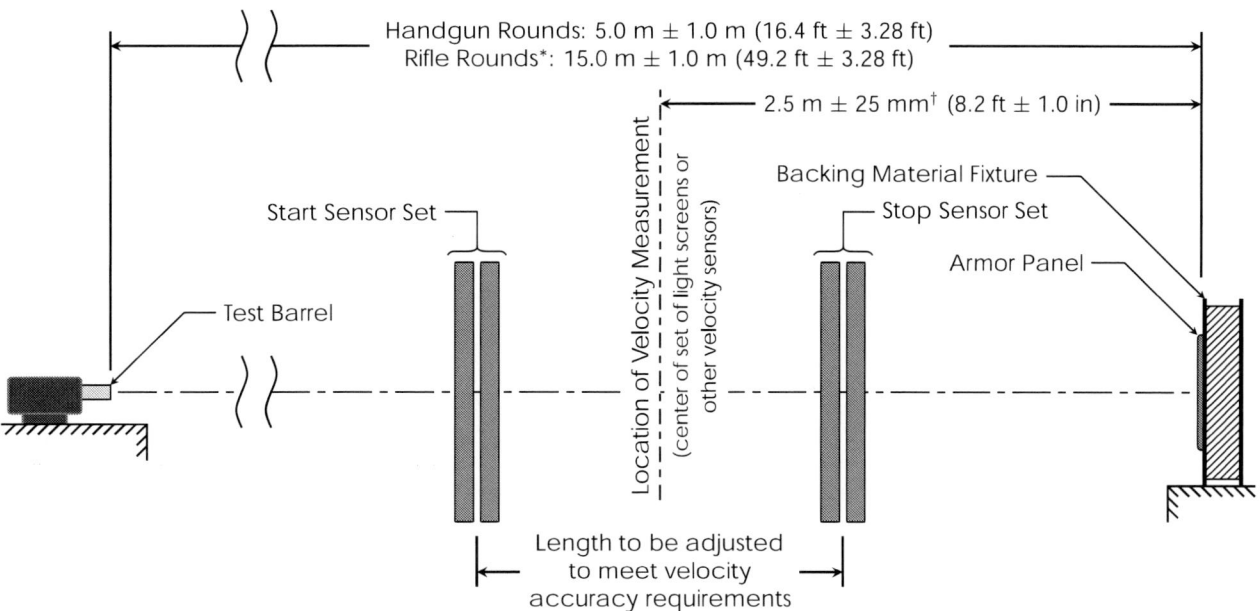

*For rifle rounds the length may be further adjusted to minimize yaw at impact; however, in such cases the yaw at impact must be experimentally shown to be less than 5° and reasonably close to minimal.
†Tolerance for 0° shots. For 30° and 45° shots the tolerance shall be + 25 mm/- 190 mm (+ 1.0 in/- 7.5 in).

Figure 8. Test range configuration

4.2.2 Test Rounds and Barrels

4.2.2.1 Handloads

Handloaded rounds will generally be necessary to achieve the required velocities specified for the P-BFS tests in table 4 of section 7. Actual velocities achieved shall be within ± 30 ft/s of the specified velocity. The bullets shall be as specified in appendix A.

4.2.2.2 Test Barrels

The test barrels shall be ANSI/SAAMI unvented velocity test barrels (see section 8, References [2], [3], and [4]). No firearms will be used. The rifling of all barrels shall meet the ANSI/SAAMI specifications for that caliber round. Barrels shall not be shorter than the ANSI/SAAMI specifications; however, longer barrels may be used when necessary for achieving the required bullet velocity. Barrels with nonstandard chambers may also be used to allow the velocities required for BL testing.

4.2.2.3 Test Barrel Fixtures

The test barrels shall be mounted in an ANSI/SAAMI compatible universal receiver or an equivalent mounting fixture. The receiver/mount will be attached to a table or other fixture having sufficient mass and restraint to ensure accurate targeting of repetitively fired rounds.

4.2.3 Velocity Measurement Equipment

Test round velocities shall be determined using at least two independent sets of instrumentation. Fair velocity measurements are individual velocity measurements within 3 m/s (10 ft/s) of each other. The velocity recorded shall be the arithmetic mean of all fair velocity measurements. The velocity measurement instrumentation shall have a combined uncertainty of less than 1.0 m/s (3.3 ft/s).

4.2.3.1 Configuration

The velocity shall be measured at 2.5 m ± 0.025 m (8.2 ft ± 1.0 in) from the front face of the backing material. When a chronograph is used in conjunction with trigger screens, the screens shall be centered at 2.5 m ± 0.025 m (8.2 ft ± 1.0 in) in front of the front surface of the backing material. For angled shots only, the screens may be centered at 2.5 m + 0.025 m / - 0.190 m (8.2 ft + 1.0 in / - 7.5 in) in front of the front surface of the backing material. Such screens shall be securely mounted to maintain their required position and spacing.

For angled shots, the positioning tolerance for the trigger screens is relaxed to allow for backing material supports that pivot the backing material around a fixed point such that the range length will decrease slightly for angled shots. All instrumentation should be positioned with the armor sample and backing material perpendicular to the line of fire and does not need to be repositioned for the angled shots.

4.2.4 Armor Submersion Equipment

The armor submersion equipment shall consist of a water bath sufficiently sized to allow at least one armor panel of the largest template size defined in appendix C to hang vertically, without any folds or bends, with the top edge of the armor at least 100 mm (3.9 in) below the surface of the water, and with at least 50 mm (2.0 in) clearance around the panel.

The water in the bath shall be clean and shall be either potable tap or demineralized water. The water shall be replaced anytime there are visible impurities in the water. The water temperature shall be 21 °C +2.9 °C/-5.8 °C (70 °F +5 °F/-10 °F).

4.2.5 Armor Backing Material

4.2.5.1 Backing Material Fixture

The inside dimensions of the backing material fixture shall be 610 mm x 610 mm with a depth of 140 mm (24.0 in x 24.0 in x 5.5 in). The tolerance on all dimensions is ± 2 mm (± 0.08 in).

The back of the fixture shall be removable and shall be constructed of 19.1 mm (0.75 in) thick wood or plywood.

4.2.5.2 Fixture Construction

The sides of the fixture shall be constructed of rigid wood or metal, preferably with a metal front edge to reliably guide the preparation of the flat front surface of the backing material.

4.2.5.3 Backing Material

In the interest of inter-laboratory conformity, Roma Plastilina No.1[1] oil-based modeling clay shall be used as the backing material. The backing material shall be replaced on an annual basis or more frequently if necessary.

4.2.5.4 Backing Material Surface Preparation

The clay in each backing material fixture shall be manipulated to produce a block free of voids, with a smooth front surface for the accurate and consistent measurement of depression depths. The front surface of the backing material shall be even with the reference surface plane defined by the fixture edges. Striking devices of sufficient length shall be used to ensure the reference surface is established using the edges of the clay block fixture as index points.

For nonplanar body armor, backing material shall be built up to conform to the shape of the nonplanar armor. Calibration drop testing shall not be performed in the built-up area. The

---

[1] The use of brand names in this standard does not constitute endorsement by the U.S. Department of Justice; Office of Justice Programs; National Institute of Justice; U.S. Department of Commerce; National Institute of Standards and Technology; Office of Law Enforcement Standards; or any other agency of the United States Federal Government, nor does it imply that the product is best suited for other applications.

built-up backing material shall be conditioned to the same temperature as the backing material in the fixture.

During preparation and post-test repair, effort shall be made to locate and remove any debris, including but not limited to bullet fragments and armor components introduced into the block during testing.

4.2.5.5 Backing Material Conditioning

The clay backing material shall be conditioned in its fixture using a heated chamber or enclosure. Actual conditioning temperature and recovery time between uses will be determined by the results of the validation drop test described in section 4.2.5.6.

Additional clay, conditioned to the same initial temperature as the fixture, shall be used to fill voids and restore the front surface of the backing material as needed.

4.2.5.6 Backing Material Consistency Validation

Drop-test validation of the backing material shall be accomplished before every six shot P-BFS test sequence and before each 12 shot BL test sequence. During P-BFS testing, a post-test drop series shall also be performed immediately following the last test sequence before the backing material is returned to the conditioning chamber. Failure to meet drop-test depth requirements will result in the invalidation of all shot series since the last drop-test with acceptable depths of indentation, and will require that a new conditioned and drop test validated backing material fixture be used. Validation shall be accomplished using the equipment and techniques specified below:

(a) Drop weight: Steel sphere.[2]
(b) Drop weight size: 63.5 mm ± 0.05 mm (2.5 in ± 0.001 in) in diameter.
(c) Drop weight mass: 1043 g ± 5 g (2.29 lb ± 0.01 lb).
(d) Drop height: 2.0 m (6.56 ft).
(e) Drop spacing: Minimum of 76 mm (3.0 in) from fixture edge to indent edge and a minimum of 152 mm (6.0 in) between indent centers.

Each validation drop will consist of a free release, targeted fall of the steel sphere onto the conditioned backing material. It is recommended that an aiming device, such as a pointing laser, be used to indicate the intended drop point on the backing material fixture. Five drops will be completed. The pretest and post-test drop positions will be located according to figure 9 and item (e) above.

The arithmetic mean of the five indentation depth measurements shall be 19 mm ± 2 mm (0.748 in ± 0.08 in). In addition, no indentation shall be greater than 22 mm (0.866 in) or less than 16 mm (0.630 in).

---

[2] A sphere, reference P/N 3606, supplied by Salem Specialty Ball Co., Inc., P.O. Box 145, West Simsbury, CT 06092, has been found to be satisfactory, although any steel sphere meeting the requirements listed in this section is acceptable.

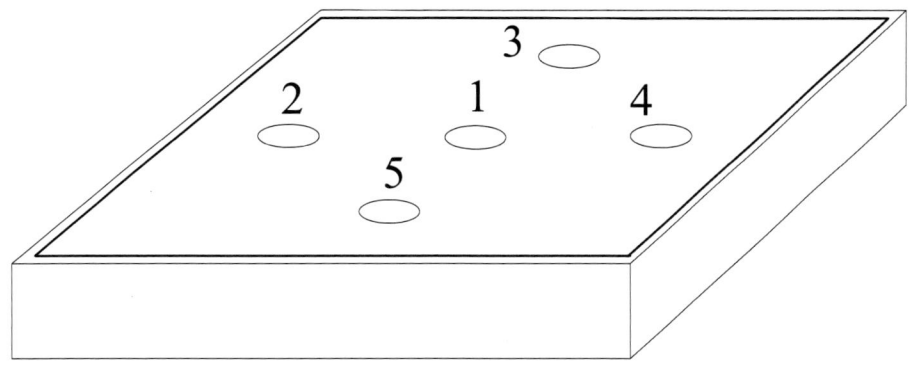

Figure 9. General pretest drop locations

Depth of indentation measurements shall utilize measurement devices (± 1 mm accuracy) indexed to the undisturbed reference surface or the edges of the fixture, establishing the reference plane across the diameter of the indentation. Depending on the chosen measurement method, the backing material may be struck flush with the edges of the fixture to reestablish the reference surface prior to measuring each depth of indentation.

Backing material temperatures shall be measured using a thermometer or thermocouple with a measurement accuracy of 0.5 °C (0.9 °F) or better. Temperature readings will be taken prior to pre and post-test drop testing at a minimum of 254 mm x 254 mm (10.0 in x 10.0 in) from any two fixture edges at a minimum depth of 25 mm (1.0 in) and a maximum depth of 51 mm (2.0 in) from the backing material surface.

A newly conditioned and drop-test validated backing material fixture shall be used for each threat specific, 24 shot series of firings or more frequently when drop test results dictate.

FLEXIBLE ARMOR CONDITIONING PROTOCOL

5.1 Purpose and Scope

This protocol applies to all Type IIA, II, and IIIA systems and to the flexible portion of Type III and Type IV armor systems. After being subjected to the conditioning protocol, flexible armor performance will be verified with ballistic testing as described in section 7.

This protocol is designed to subject test armors to conditions that are intended to provide some indication of the armor's ability to maintain ballistic performance after being exposed to conditions of heat, moisture, and mechanical wear. This protocol will not predict the service life of the vest nor does it simulate an exact period of time in the field.

5.2 Pretest Parameters

5.2.1 Storage of Armors

To allow for equilibration, store the test samples at a temperature of 25 °C ± 10 °C (77 °F ± 18 °F) with a relative humidity of 20 % to 50 % for at least 24 h prior to starting the armor conditioning protocol. This may be achieved in a controlled laboratory environment, or if conditions in the laboratory deviate from those specified, a chamber may be used to create these conditions.

5.2.2 Pretest Calibrations

Prior to and following each conditioning protocol, the accuracy of all instruments and test equipment used to control or monitor the test parameters shall be verified. The calibration intervals must meet ANSI or ISO guidelines for good laboratory practices.

5.2.3 Test Conditions

5.2.3.1 Air Temperature

Keep the air temperature uniform, both inside the conditioning chamber and in the storage environment. Verify that the air temperature is uniform by using verification sensors to ensure that the air temperature is within ± 2 °C (± 3.6 °F) of the required temperature. Storage temperatures are given in section 5.2.1.

The test temperature shall be 65 °C (149 °F).

5.2.3.2 Relative Humidity

Keep the relative humidity uniform and noncondensing, both inside the test chamber and in the storage environment. Verify that the relative humidity is uniform by using verification sensors to ensure that the relative humidity is within ± 5 % of the required relative humidity. The storage relative humidity is given in section 5.2.1.

The test relative humidity shall be 80 %.

5.2.3.3 Tumbling

The drum rotation rate shall be 5.0 rpm ± 1.0 rpm.

The drum shall be rotated through 72,000 ± 1,500 complete rotations during the test. To ensure that the proper number of complete rotations is obtained, a counter or totalizer shall be used to record the total number of rotations. The drum rotation rate may be varied within the given tolerance to achieve the necessary number of rotations, as long as the rate does not exceed the above specified tolerance.

5.2.3.4 Test duration

The test shall last a total of 10 d ± 1 h.

5.3 Laboratory Configuration and Test Equipment

5.3.1 General Parameters

5.3.1.1 Facility Design

Use a stand-alone or walk-in temperature and humidity chamber with the tumbling accessory inside.

5.3.1.2 Monitoring Conditioning Chamber Parameters

To provide proof of parameter level maintenance, the laboratory must keep a manually or electronically produced log of parameter levels. Exact parameter monitoring intervals and exact methods of recording parameter levels may vary for different laboratories. The technology involved in recording parameter levels may involve visual checks at prescribed intervals, real time continuous recording such as a circular chart, periodic recording on a device such as a data logger, or other techniques as appropriate for each individual laboratory.

5.3.1.3 Humidity Generation

Use steam or water injection to create the relative humidity within the test chamber as appropriate for the device. Set up a system to drain and discard any condensate developed within the chamber during the test.

5.3.1.3.1 Water Purity

It is essential that the water used for humidity tests not unfairly impose contaminants on the test samples. Chemicals commonly found in commercial water supplies, such as chlorine, as well as nonneutral pH, can cause unintended corrosive effects to materials. It is highly recommended that the water used for this test be relatively free from impurities and chemicals and have a pH in the range of 6.5 to 7.2. **NOTE**: A water resistivity of no less than 150,000 ohm cm is recommended. This can be produced using distillation, demineralization, reverse osmosis, or deionization. Many humidity chambers have a built in deionization system.

5.3.1.4 Tumbling Generation

Use a tumbling device to generate tumbling action during the protocol.

The tumbler drum shall have an internal diameter of 832 mm ± 6 mm (32 3/4 in ± 1/4 in) and an internal depth of 651 mm ± 6 mm (25 5/8 in ± 1/4 in). The tumbler drum shall have four fins (ribs) running the full depth of the drum and spaced at 90º intervals about the circumference. Each fin shall be 114 mm ± 3 mm (4 1/2 in ± 1/8 in) high. The top edge of the fins shall be rounded with a minimum diameter of 19 mm ± 3 mm (3/4 in ± 1/8 in), and the width of the fin at the top shall be 19 mm ± 3 mm (3/4 in ± 1/8 in). The fins may be either straight or tapered. The base of the fin shall not be thinner than 19 mm (3/4 in) and shall not be thicker than 76 mm (3.0 in). The inside of the drum shall be smooth, with no sharp edges to catch, tear, or abrade the armor samples. The drum surface may be perforated. The drum shall have sufficient openings to allow air flow such that the air within the drum remains within the specified tolerances.

The tumbling apparatus should be placed in a chamber capable of generating temperature and humidity during the conditioning protocol. The tumbler must maintain tolerances of temperature, humidity, and revolutions per minute specified in section 5.2.3. Verify the apparatus maintains conditions within the tolerances specified at periodic intervals throughout each cycle of the conditioning protocol. The tumbler and chamber must be arranged such that if the chamber goes out of the tolerance conditions, the tumbler rotation will cease until the chamber returns to its in-tolerance condition.

5.3.1.5 Contamination Prevention

Do not bring any material other than water into physical contact with the test armors. Do not introduce any material other than water into the chamber.

Care shall be taken to prevent condensation within the humidity chamber from coming in contact with the armor samples.

5.3.2 Controls

(a) Ensure the test chamber includes measurement and recording device(s), separate from the chamber controllers.
(b) Unless otherwise specified, make continuous analog temperature and relative humidity measurements during the test. Conduct digital measurements at intervals of 10 min. or less.
(c) Use only instrumentation with the selected conditioning chamber that meets the accuracies, tolerances, etc., described herein.

5.3.3 Test Interruption

Analyze any interruption carefully. If the decision is made to continue the protocol from a point of interruption or to add additional time onto the protocol, and a failure occurs, it is essential to be able to determine the effects of the protocol interruption. The flow chart shown in figure 10 will assist in the determination of how to proceed if an interruption occurs.

5.3.4 Procedure

This procedure consists of a 24-h acclimation period to ensure all armor samples start with the same conditions followed by a 10-d period of temperature, humidity, and mechanical wear exposure. This procedure shall be performed on only one compliance test group at a time.

[1] When a compliance test group is received, place the samples in a controlled laboratory environment for at least 24 h prior to beginning the test as specified in section 5.2.1.
[2] Place the compliance test group inside of the tumbler at the specified conditions of temperature and relative humidity (see section 5.2.3).
[3] Program the tumbler to rotate as specified in section 5.2.3.3.
[4] Expose the armors to the specified conditions for the period specified in section 5.2.3.4.
[5] Return armors to pretest conditions (see section 5.2.1).
[6] Perform a thorough visual examination of each test sample and document any change in physical appearance resulting from exposure.
[7] Armors shall remain at pretest conditions (see section 5.2.1) for a minimum of 12 h before transporting them to a different facility or before beginning ballistic testing.

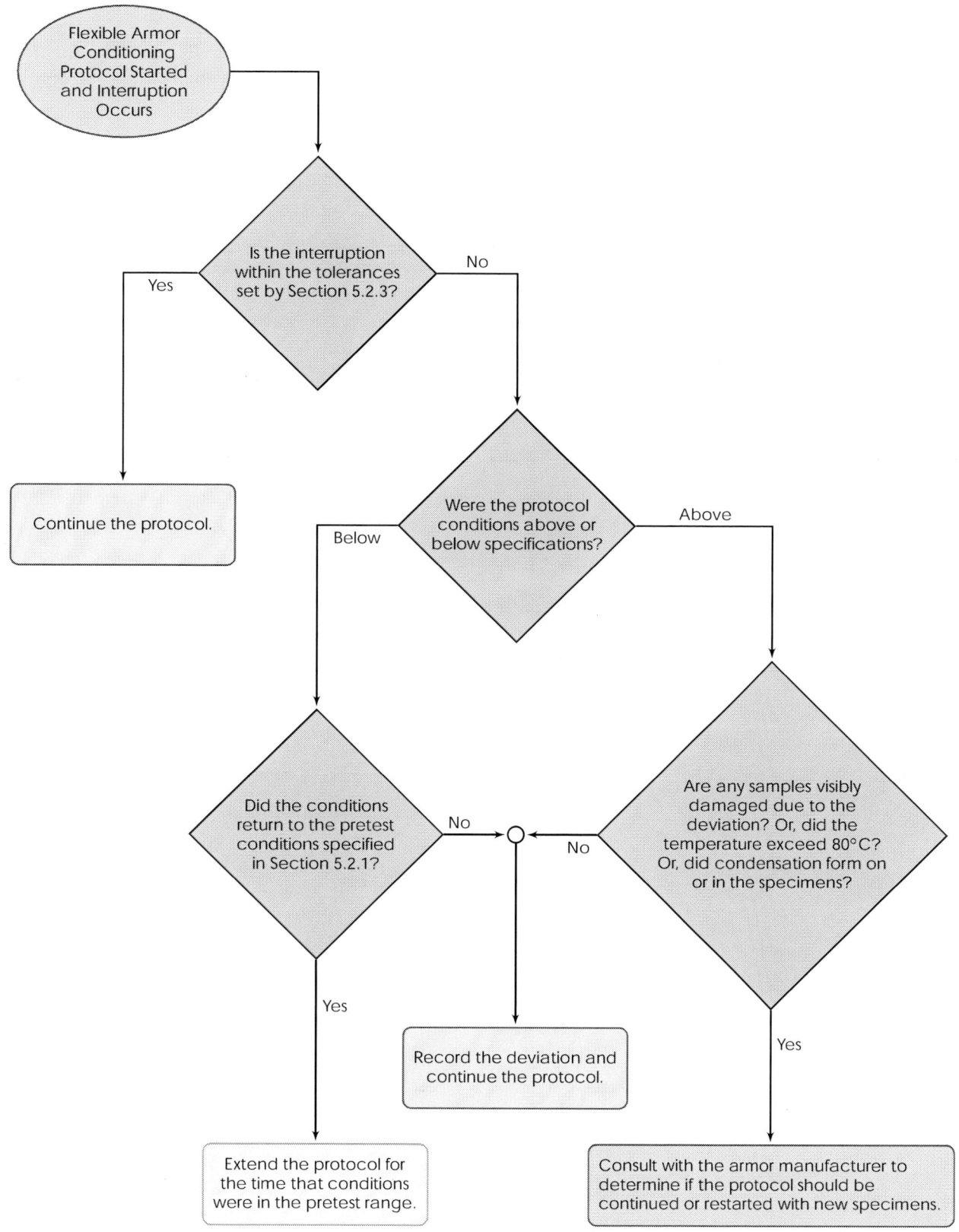

Figure 10. Flexible armor conditioning test interruption flow chart

HARD ARMOR CONDITIONING PROTOCOL

6.1 Purpose and Scope

This section is designed to subject hard armors or plate inserts to conditions that are intended to provide some indication of armor's ability to maintain ballistic performance after being exposed to conditions of heat, moisture, and mechanical wear. This protocol is intended to apply only to rigid systems. After this armor conditioning protocol, the ballistic performance will be verified with ballistic testing as described in section 7.

This protocol will not predict the service life of the armor nor does it simulate an exact period of time in the field.

This protocol consists of a four step process: (1) a 24-h acclimation period to ensure all items at any intended climatic test location will start with the same conditions, (2) a 10-d uniform thermal exposure, (3) a thermal cycle exposure, and (4) mechanical durability testing.

6.2 Pretest Parameters

6.2.1 Storage of Armors

To allow for equilibration of armors, store the test armors at a temperature of 25 °C ± 10 °C (77 °F ± 18 °F) and relative humidity of 20 % to 50 % for at least 24 h prior to starting the armor conditioning protocol.

6.2.2 Pretest Calibrations

Prior to and following each armor conditioning protocol, the accuracy of all instruments and test equipment used to control or monitor the test parameters shall be verified. The calibration intervals must meet ANSI or ISO guidelines for good laboratory practices.

6.2.3 Test Conditions

6.2.3.1 Air Temperature

Keep the air temperature uniform, both inside the test chamber and in the storage environment. Verify that the air temperature is uniform by using verification sensors to ensure that the air temperature is within ± 2 °C (± 3.6 °F) of the required temperature. Storage temperatures are given in section 6.2.1. Test temperatures are given in table 2 for uniform thermal exposure testing and in table 3 for the thermal cycle testing.

Table 2. Uniform thermal exposure conditions

Time	Temperature	Relative Humidity
10 d	65 °C (149 °F)	80 %

Table 3. Thermal cycle testing conditions

Time (hr)	Temperature (°C)	Relative humidity (%)
2	25	50
2	15	N/A
2	5	N/A
2	-5	N/A
2	-15	N/A
2	0	N/A
2	15	N/A
2	30	50
2	45	50
2	60	50
2	75	50
2	90	50

6.2.3.2 Relative Humidity

Keep the relative humidity uniform, both inside the test chamber and in the storage environment. Verify that the relative humidity is uniform by using verification sensors to ensure that the relative humidity is within ± 5 %. Storage relative humidities are given in section 6.2.1. Test relative humidities are given in table 2 for uniform thermal exposure and in table 3 for the thermal cycle testing.

6.2.3.3 Armor Drop Test

A flat hardened concrete surface with a thickness of at least 76.2 mm (3.0 in) and a mass much greater than that expected from the weighted plate must be available for mechanical durability testing.

6.2.3.4 Test Duration

The uniform thermal exposure test shall last a total of 10 d. The thermal cycle test will last a total of 1 d.

6.2.3.5 Monitoring Conditioning Chamber Parameters

(a) To provide proof of parameter level maintenance, keep a manually or electronically produced log of parameter levels. Exact parameter monitoring intervals and exact methods of recording parameter levels may vary for different laboratories.
(b) The technology involved in recording parameter levels may involve visual checks at prescribed intervals, real time continuous recording such as a circular chart, periodic recording on a device such as a data logger, or other techniques as appropriate for each individual laboratory.

6.3 Laboratory Configuration and Test Equipment

6.3.1 General Parameters

6.3.1.1 Facility Design

Use a stand-alone or walk-in humidity chamber with racks positioned such that the plates can be mounted in a vertical orientation (the orientation in which they will be used).

6.3.1.2 Test Sensors and Measurements

The laboratory must have a method of measuring and recording the temperature and relative humidity inside the chamber at intervals not less than once every 10 min. This can be accomplished by using monitoring software on the humidity chamber, an array of sensors, a chart recorder, or other appropriate methods.

6.3.1.3 Humidity Generation

Use steam or water injection to create the relative humidity within the test chamber as appropriate for the device. Set up a system to drain and discard any condensate developed within the chamber during the test.

6.3.1.3.1 Water Purity

It is essential that the water used for humidity tests not unfairly impose contaminants on the test armors. Chemicals commonly found in commercial water supplies such as chlorine as well as nonneutral pH can cause unintended corrosive effects to materials. It is highly recommended that the water used for this test be relatively clean of impurities and chemicals, and have a pH in the range of 6.5 to 7.2 at 25 °C at the time of test. **NOTE**: A water resistivity of no less than 150,000 ohm cm is recommended. This can be produced using distillation, demineralization, reverse osmosis, or deionization. Many humidity chambers have a built-in deionization system.

6.3.2 Controls

(a) Ensure the test chamber includes measurement and recording device(s), separate from the chamber controllers.
(b) Unless otherwise specified, make continuous analog temperature and relative humidity measurements during the test. Conduct digital measurements at intervals of 10 min or less.
(c) Use only instrumentation with the selected test chamber that meets the accuracies, tolerances, etc., described herein.

6.3.3 Test Interruption

Analyze any interruption carefully. If the decision is made to continue the protocol from a point of interruption or to add additional time onto the protocol, and a failure occurs, it is essential to be able to determine the effects of the test interruption. The flow chart shown in figure 11 will assist in the determination of how to proceed if an interruption occurs.

6.3.4 Conditioning Procedure

The conditioning procedure is detailed below:

[1] When samples are received, place them in a controlled laboratory environment for at least 24 h prior to beginning the test.
[2] Place the armors inside of the chamber at the specified conditions of temperature and relative humidity.
[3] Expose the armors to the conditions specified in table 2 for 10 d.
[4] Perform a thorough visual examination of the test item and document any change in physical appearance resulting from exposure.
[5] Expose the armors to the conditions specified in table 3 for 24 h.
[6] Perform a thorough visual examination of the test item and document any change in physical appearance resulting from exposure.
[7] Perform drop testing on the armors by attaching them with a strap, belt, or other nonobstructive retaining device to the front surface of 10.0 lb of P-BFS backing material. The backing material clay will be contoured to the back surface of the plate. If the plate is part of an in conjunction system, the plate shall be backed for drop testing by the in conjunction flexible armor design intended to be sold with the plate. This flexible armor design must be supplied by the manufacturer in template size NIJ–C–2. This flexible armor is placed between the armor and the weighted object as shown in figure 12. If the plate is intended to be sold as a stand-alone item, no flexible armor backing will be used in drop testing. A fixture similar to that shown in Figure 12 has been shown to provide accurate and reproducible results. For further details and specifications on this fixture, see the publication entitled, *Description of Fixture for Hard Armor Drop Testing According to NIJ Standard-0101.06*. At a minimum free fall height of 48.0 in (90° from horizontal) from the flat hardened surface and with the strike face of the armor pointing down, drop the weighted fixture twice. Impacts must occur at the center of the face (not at an edge).

Following conditioning testing, the armor shall be subjected to further testing according to section 7.

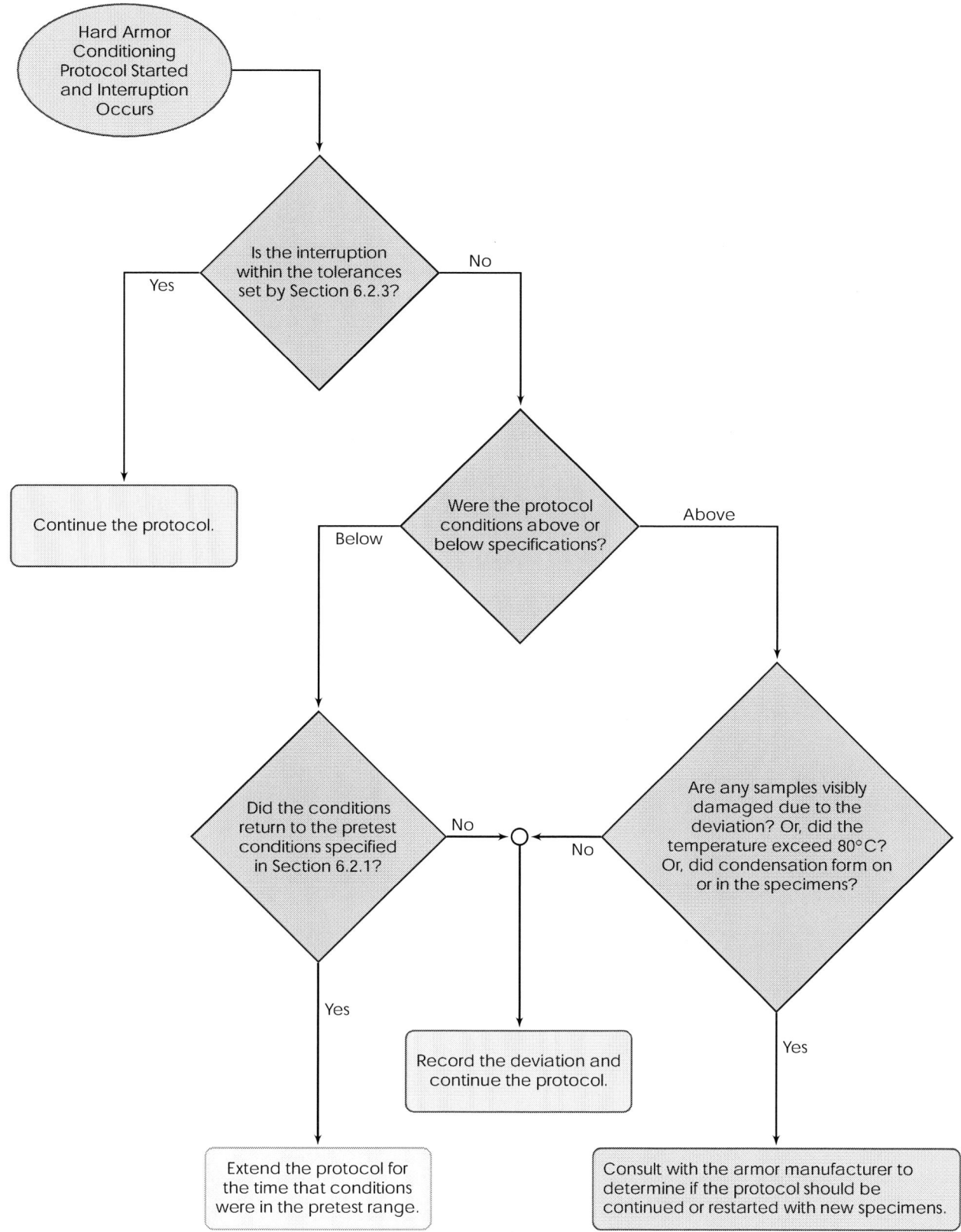

Figure 11. Test interruption flowchart for hard armor

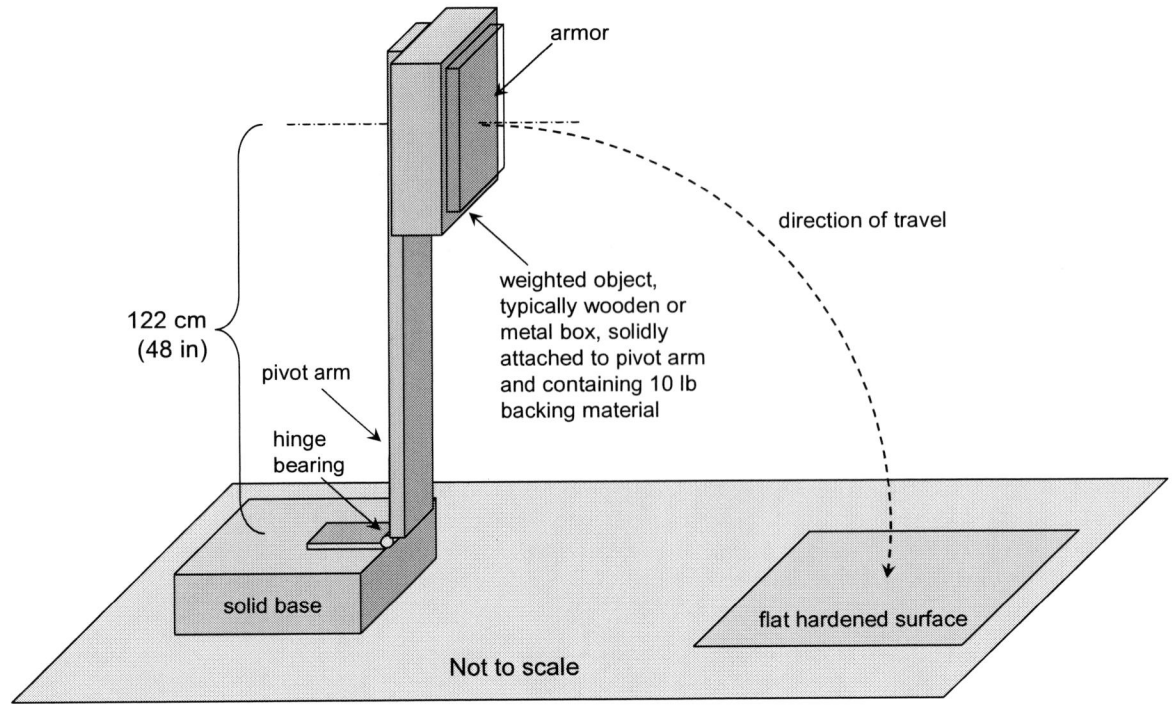

Figure 12. Proposed mechanical durability test apparatus [3]

---

[3] From *Purchase Description Personal Armor, Enhanced Small Arms Protective Insert* CO/PD 04–19A 28 February 2005.

BALLISTIC TEST METHODS

7.1 Purpose and Scope

This section specifies the methods and performance requirements for ballistic testing of body armor and includes the formal test procedures for the Perforation-Backface Signature (P-BFS) and baseline Ballistic Limit (BL) tests. The first test series is P-BFS testing and requires the armor to demonstrate consistent ballistic resistance to both perforation and excessive blunt force trauma. The second test series is BL testing and is designed to statistically estimate perforation performance. In addition, for flexible vests and jackets, a portion of the sample set will be subjected to the flexible armor conditioning protocol described in section 5. For hard armor and plate inserts, the entire sample set will be subjected to the hard armor conditioning protocol described in section 6. The conditioned samples will be subjected to ballistic testing.

7.2 Test Order

The ballistic test series may be performed in any order, at the discretion of the manufacturer.

7.3 Workmanship Examination

All armor samples received for compliance testing shall be inspected before testing, and any deficiencies shall be noted. Samples that do meet the requirements of section 4.1 shall be reported to both the armor manufacturer and the Compliance Testing Program (CTP) before testing commences. Documentary photographs shall be taken for use in deficiency notification reporting.

7.3.1 Armor Carriers and Ballistic Panel Covers

All armor sample carriers and ballistic panel covers received for compliance testing will be inspected for damage, material flaws, or poor workmanship as defined in section 4.1.4. Documentary photographs shall be taken for use in deficiency notification reporting.

7.3.2 Ballistic Panels

<u>Pretest</u> – Before testing, all armor sample ballistic panels and inserts received for compliance testing will be individually inspected for damage, material flaws, or poor workmanship as defined in section 4.1.4.

<u>Post-Test</u> - Each sample's ballistic components will be inspected immediately after testing to verify their construction details (layers, weave, stitching, material, etc.).

7.3.3 Label Examination

The complete armor sample and each part (carrier and ballistic panels) will be examined for conformance to the labeling requirements of section 4.1.5.

7.4 Sampling

Test samples will be selected according to the requirements of section 4.1.1.2.

7.5 Sample Acclimation

All armor samples received for compliance testing shall be stored and acclimated for a minimum of 24 h at ambient range conditions.

7.5.1 Inserts

All armor samples will be tested in their final design and end use configuration, including all system components (e.g., carriers and straps), with the following exceptions. Armors with removable carriers that may be manufactured with more than one carrier design shall be tested with the minimal carrier defined in section 4.1.6. Except for plate inserts being tested as in conjunction armors, all removable inserts or trauma packs shall be removed from the armor before acclimation and shall not be tested.

7.6 Fair Hit Requirements for All Ballistic Tests

A test shot shall be considered a fair hit if it impacts the armor panel at an angle of incidence no greater than ± 5° from the intended angle of incidence, no closer to the edge of the ballistic panel than the minimum *shot-to-edge distance,* and no closer to a prior hit than the minimum *shot-to-shot distance.*

In addition, for a P-BFS test shot to be considered a fair hit, it must meet the impact location requirements of section 7.8.1, and the measured velocity must either: (1) be within ± 9.1 m/s (30 ft/s) of the reference velocity for the specified bullet; (2) be *less than* the minimum velocity and *produce a perforation or an excessive BFS*; or (3) be *greater than* the maximum velocity and *not produce a perforation or an excessive BFS*.

7.6.1 Minimum Shot-to-Edge Distance

The armor manufacturer shall define the acceptable minimum shot-to-edge distance for each armor model and each threat that will be tested against the armor. For armor types subjected to a single threat and for the lighter weight threat round when two threats are specified, the minimum shot-to-edge distance shall not be greater than 51 mm (2.0 in). For the heavier threat round when two threats are specified, the minimum shot-to-edge distance shall not be greater than 76 mm (3.0 in). Table 4 indicates which rounds are lighter weight and heavier weight.

7.6.2 Minimum Shot-to-Shot Distance

The minimum shot-to-shot distance shall be 51 mm (2.0 in).

This distance may be decreased at the request of the armor manufacturer.

7.7 Backing Material Preparation and Sample Mounting for All Ballistic Tests

7.7.1 Backing Material Fixture Preparation

The backing material fixture shall meet the requirements of section 4.2.5.

The clay in each backing material fixture will be manipulated to produce a block free of voids and with a smooth, flat front surface for the accurate and consistent measurement of depression depths. The front surface of the backing material shall be even with the reference surface plane defined by the fixture edges. Striking devices of sufficient length shall be used to ensure the reference surface is established using the edges of the fixture as index points.

For BL tests only, the removable back shall be removed from the backing material fixture.

7.7.2 Mounting Armor for Ballistic Testing

7.7.2.1 Strapping

The armor panel shall be positioned on the backing material such that the point of impact, projected through the armor onto the surface of the backing material, is no closer that 106 mm (4.2 in) from the edge of the backing material fixture.

Armor samples or panels shall be held in contact with the backing material and secured to the backing material fixture using mounting straps that conform to one of the following two options:

(a) The default strapping method uses 51 mm (2.0 in) wide elastic straps, held together using hook-and-loop fasteners. If these straps are used, two vertical and three horizontal straps shall be positioned such that they do not interfere with the impact points on the armor panel. Figure 13, diagram A shows a typical example of strapping using this method. For larger armor sizes, the backing material fixture is not large enough to accommodate the entire armor. For these cases, extension panels shall be added to the sides of the fixture, as shown in figure 13, diagram B, such that the combined backing material, backing material fixture, and extensions form a planar surface at least 1016 mm (40.0 in) wide by 610 mm (24.0 in) high. The extensions may be part of the backing material support fixture, and this configuration may be used for smaller armors.
(b) The alternative strapping method applies to armor panels that normally would have strapping as an integral part of their construction. This strapping method requires the armor manufacturer to supply armor panels with extended strapping devices to allow the armor, as a unit, to be mounted on the backing material fixture. Figure 13, diagram C shows a typical example of strapping using this method.

The laboratory shall record which strapping arrangement was used.

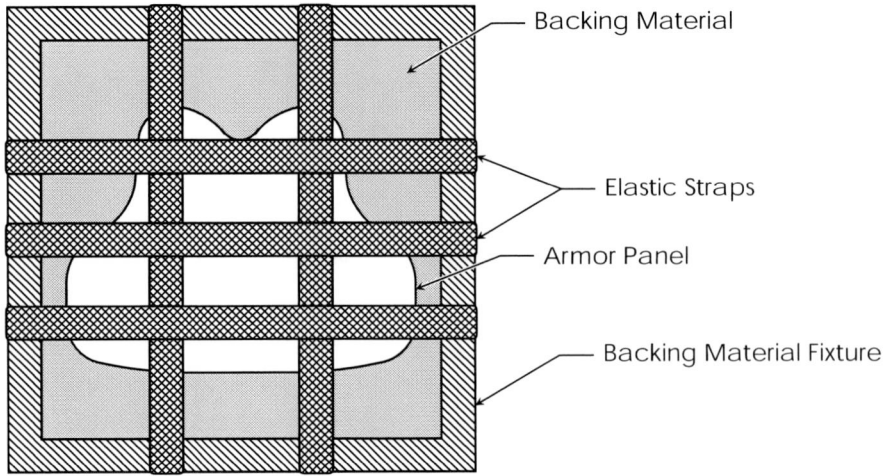
A. Standard Strapping Arrangement for Smaller Samples

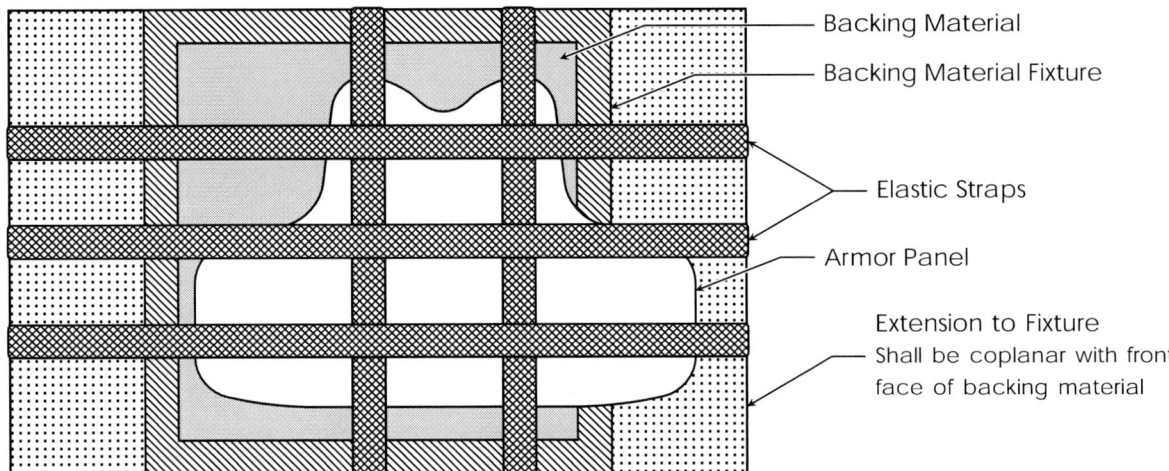
B. Standard Strapping Arrangement for Larger Samples

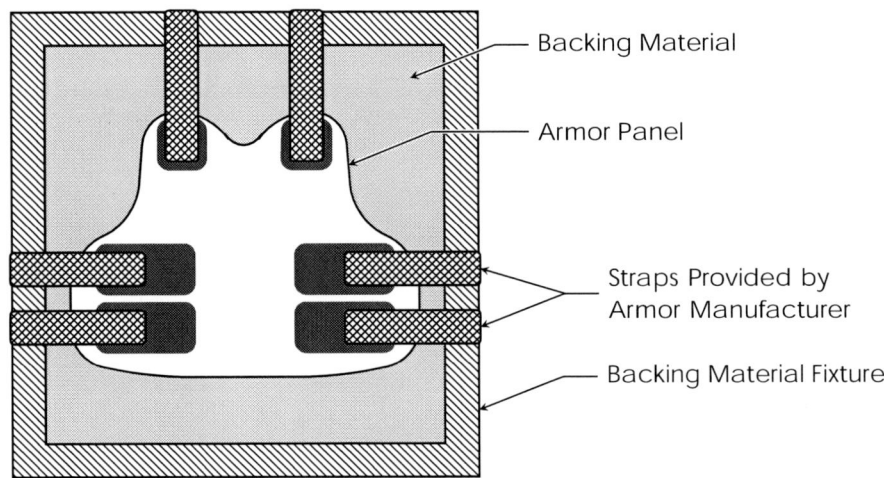
C. Special Strapping Arrangement Using the Armor's Strap Attachments

Figure 13. Acceptable strapping methods

7.7.2.2 Remounting of Armor

Once the armor panel has been subjected to any ballistic tests, no effort shall be made to recover any bullets trapped in the armor panel, and no effort shall be made to fill in any BFS indentations in the test fixture backing material. When necessary, the backing material shall be struck to return the surface to a flat configuration; however, the BFS indentation should not be repaired. The armor panel shall be manipulated by hand so that any deformations in the armor are smoothed out. Also, the armor panel shall be repositioned on the backing material such that the next impact will not occur over any BFS indentation and such that the panel is supported by smooth clay backing material for a distance of no less than 76 mm (3.0 in) in all directions around the next shot location.

7.7.2.3 Backing Material Fixture Positioning

The backing material fixture shall be positioned to ensure proper bullet impact placement and angle of incidence of the test round. For any shots requiring a nonzero angle of incidence, the backing material fixture shall be positioned so that the bullet line of flight is directed toward the vertical centerline of the armor panel at the point of impact. An exception will be made for armors where shots directed in a different orientation will likely be more problematic for the armor to stop. In these cases, for shots requiring a nonzero angle of incidence, the armor shall be oriented such that the shot will exploit the perceived weakness.

7.8 Perforation and Backface Signature Test (P-BFS)

All armor models submitted for compliance testing will undergo P-BFS testing using the threat rounds specified in section 2 and table 4. These impact tests measure three BFS indentions and demonstrate the armor's resistance to perforation. This test series requires the use of a plastically deforming witness media (clay backing material) held in direct contact with the back surface of the armor panel. This configuration is used to capture and measure the BFS depression produced in the backing material during nonperforating threat round impacts.

The number of armor samples defined in section 4.1 shall be subjected to P-BFS tests in accordance with the following procedures.

7.8.1 Shot Location Marking

Clearly mark the shot locations directly on each sample according to the following criteria.

Flexible Vests and Jackets: All flexible vests and jackets shall be tested with six shots in the approximate pattern shown in figure 14. Shots 1, 2, and 3 shall meet the shot-to-edge distance requirements, but they shall not be located more than the minimum shot-to-edge distance plus 19 mm (0.75 in) from the edge of the panel. Shots 4, 5, and 6 shall meet the shot-to-shot distance requirements, but all three shots shall be located within a 100 mm (3.94 in) diameter circle. For panels with sufficient area, the locations of shots 4, 5, and 6 shall be randomly moved to different areas on different panels. When there are no discontinuities or apparent weaknesses in the armor as considered below, refer to guidance from the CTP on the specific locations for shots 4, 5, and 6 on each panel.

For the case of armor samples sized to Template NIJ–C–1 (smallest template) when tested with threat 2 only, the panel may be subjected to only five shots if space limitations prevent taking six shots. In this case, shots 1 through 5 shall be taken, but shot 6 may be skipped. In this case only, the total number of shots required for a complete P-BFS test series shall be 66 shots per test threat.

For armors where the construction and material thickness vary across the panel, the locations of shots 4, 5, and 6 shall be adjusted to exploit the weakest portion of the armor.

For armors that have folds, seams, or other discontinuities (such as the bust cups of some female armors or the closures of front closing vests), additional shots shall be fired so that at least one shot impacts each fold, seam, or discontinuity. If a single fold, seam, or discontinuity extends more than one half the width or height of the armor (such as closure of a front closing vest), at least two shots shall impact that discontinuity. For small armors with limited shot area, the locations of shots 4, 5, and 6 may be varied such that one of these shots impacts the discontinuity.

Hard Armors and Plate Inserts: All hard armors and plate inserts shall be tested with the appropriate number of shots as defined in section 4.1.2. The shots shall be placed on the panel in any pattern that meets the shot-to-edge and shot-to-shot spacing requirements.

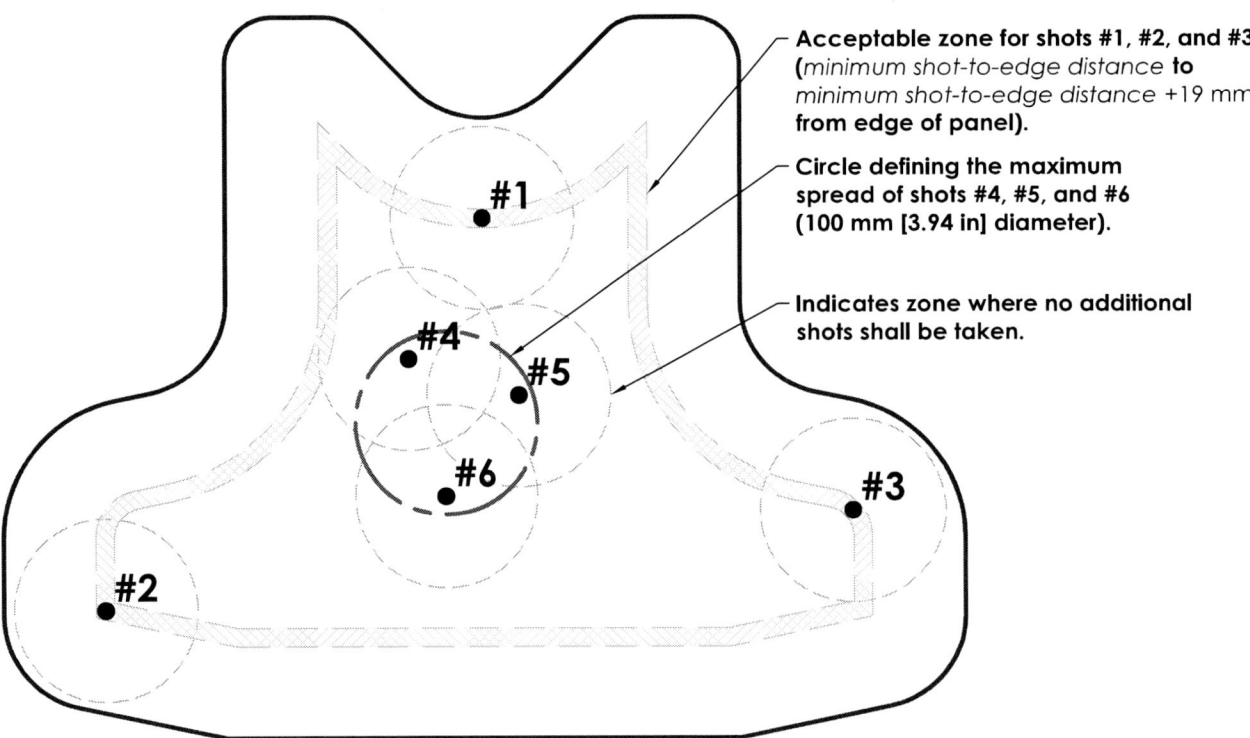

Figure 14. General armor panel impact locations (front and back)

7.8.2 Armor Submersion

New Flexible Vests and Jackets: All new flexible vests and jackets shall be submersed and tested wet. Each armor panel shall be hung vertically in a water bath meeting the requirements of section 4.2.4 for 30 min (+ 5 min/- 0 min) with the top edge of the armor positioned 100 mm ± 25 mm (3.9 in ± 1.0 in) below the water surface. For armors that are buoyant, weights shall be attached to the bottom edge of the armor with clothes pins or similar clips to allow the armor to hang vertically. After removing the panel from the water, it shall be hung vertically and allowed to dry for 10 min (+ 5 min/- 0 min) before mounting on the test fixture. All testing of the panel shall be completed within 40 min of when the panel is removed from the water.

Conditioned Flexible Vests and Jackets: All flexible vests and jackets that have been subjected to the conditioning protocol described in section 5 shall not be submersed but shall be tested dry.

Hard Armors and Plate Inserts: All hard armors and plate inserts shall be submersed and tested wet. When plate inserts are tested in conjunction with a flexible vest or jacket, the flexible component shall have previously demonstrated its full compliance with this standard at its appropriate level. Both the plate and the flexible vest or jacket shall be tested wet.

7.8.3 Test Threats for P-BFS Tests

All armors shall be tested with the appropriate threats for the armor type as defined in section 2 and listed in table 4. In addition, a manufacturer may specify additional special type threats for a particular armor model.

7.8.4 Test Duration

The duration of the ballistic test on each armor panel shall be no longer than 30 min, from the time the first shot is fired until the last shot is fired. The first shot shall be fired within 10 min. after completion of the armor submersion cycle described in section 7.8.2. Testing shall be repeated if either of these requirements is not met. All test results, including the noncompliant test series, shall be reported. Test start and stop times shall be recorded for each test series.

7.8.5 Requirements for Number of Shots and Number of Armor Samples

A complete P-BFS test for an armor type consists of the total required number of fair hits accumulated on the required number of armor samples for each of the specified test rounds, as defined in table 5 for new condition armors and table 6 for conditioned armors. Each armor panel or plate shall be subjected to the number of fair hits specified in table 5 or table 6, as appropriate, using the specified test round. The shots on each panel shall be fired in order of their location numbers, as shown in figure 14. The angles of the shots shall be as specified in table 7. For flexible vests and jackets, one shot on each panel shall have an angle of incidence of 30°, and one shot shall have an angle of incidence of 45°. The angle of incidence for location 4 on the first P-BFS panel tested shall be randomly selected as either 30° or 45°, and the other angle used at shot location 5. For all subsequent P-BFS panels of that model, the order of the two angles shall alternate.

Table 4. P-BFS performance test summary

			Test Variables				Performance Requirements				Shot Requirements				
Armor Type	Test Round	Test Bullet	Bullet Mass	Conditioned Armor Test Velocity*	New Armor Test Velocity*	Hits Per Panel at 0° Angle	Maximum BFS Depth	Hits Per Panel at 30° or 45° Angle†	Shots Per Panel	Panel Size	Panel Condition	Panels Required	Shots Required	Total Shots Required
IIA	1	9 mm FMJ RN	8.0 g (124 gr)	355 m/s (1165 ft/s)	373 m/s (1225 ft/s)	4	44 mm (1.73 in)	2	6	Large	New	4	24	144
											Conditioned	2	12	
										Small	New	4	24	
											Conditioned	2	12	
	2	.40 S&W FMJ	11.7 g (180 gr)	325 m/s (1065 ft/s)	352 m/s (1155 ft/s)	4	44 mm (1.73 in)	2	6	Large	New	4	24	
											Conditioned	2	12	
										Small	New	4	24	
											Conditioned	2	12	
II	1	9 mm FMJ RN	8.0 g (124 gr)	379 m/s (1245 ft/s)	398 m/s (1305 ft/s)	4	44 mm (1.73 in)	2	6	Large	New	4	24	144
											Conditioned	2	12	
										Small	New	4	24	
											Conditioned	2	12	
	2	.357 Magnum JSP	10.2 g (158 gr)	408 m/s (1340 ft/s)	436 m/s (1430 ft/s)	4	44 mm (1.73 in)	2	6	Large	New	4	24	
											Conditioned	2	12	
										Small	New	4	24	
											Conditioned	2	12	
IIIA	1	.357 SIG FMJ FN	8.1 g (125 gr)	430 m/s (1410 ft/s)	448 m/s (1470 ft/s)	4	44 mm (1.73 in)	2	6	Large	New	4	24	144
											Conditioned	2	12	
										Small	New	4	24	
											Conditioned	2	12	
	2	.44 Magnum SJHP	15.6 g (240 gr)	408 m/s (1340 ft/s)	436 m/s (1430 ft/s)	4	44 mm (1.73 in)	2	6	Large	New	4	24	
											Conditioned	2	12	
										Small	New	4	24	
											Conditioned	2	12	
III	1	7.62 mm NATO FMJ	9.6 g (147 gr)	847 m/s (2780 ft/s)	-	6	44 mm (1.73 in)	0	6	All	Conditioned	4	24	24
IV	1	.30 Caliber M2 AP	10.8 g (166 gr)	878 m/s (2880 ft/s)	-	1 to 6	44 mm (1.73 in)	0	1 to 6	All	Conditioned	4 to 24	24	24
Special	-	Each test threat to be specified by armor manufacturer or procuring organization.							Armor performance and shot requirements shall depend on armor type.					

*Target measurement velocity. Fair hit measurement velocities must be within ± 9.1 m/s (± 30 ft/s) of this value, as defined in Section 7.6.
†Each armor that is to be shot at angles other than 0° shall be shot once at a 30° angle and once at a 45° angle.

Table 5. Number of shots and fair hits on each size of new condition armor samples

Armor Type	Test Round (see table 4)	Required Fair Hits Per Test Round	Required Number of Complete Armor Samples	Required Number of Armor Panels or Plates	Required Fair Hits per Panel or Plate	Maximum Number of Shots Allowed per Panel or Plate	Required Total Fair Hits per Size
IIA	1	24	2	4	6	8	48
	2	24	2	4	6	8	
II	1	24	2	4	6	8	48
	2	24	2	4	6	8	
IIIA	1	24	2	4	6	8	48
	2	24	2	4	6	8	
III	1	24	4	4	6	6	24
IV	1	24	TBD	TBD	1 to 6	6	24

Table 6. Number of shots and fair hits on each size of conditioned samples

Armor Type	Test Round (see table 4)	Required Fair Hits Per Test Round	Required Number of Complete Armor Samples	Required Number of Armor Panels or Plates	Required Fair Hits per Panel or Plate	Maximum Number of Shots Allowed per Panel or Plate	Required Total Fair Hits per size
IIA	1	12	1	2	6	8	24
	2	12	1	2	6	8	
II	1	12	1	2	6	8	24
	2	12	1	2	6	8	
IIIA	1	12	1	2	6	8	24
	2	12	1	2	6	8	
III	1	24	4	4	6	6	24
IV	1	24	TBD	TBD	1 to 6	6	24

Table 7. Angle of incidence locations

Armor Type	Test Round (see table 4)	Shot Location(s) for Fair Hits at 0° Angle of Incidence	Location(s) for Fair Hits at 30° or 45° Angle of Incidence	Shot Locations for BFS Measurement(s)
IIA	1	1, 2, 3, 6	4, 5	1, 2, 3
	2	1, 2, 3, 6	4, 5	1, 2, 3
II	1	1, 2, 3, 6	4, 5	1, 2, 3
	2	1, 2, 3, 6	4, 5	1, 2, 3
IIIA	1	1, 2, 3, 6	4, 5	1, 2, 3
	2	1, 2, 3, 6	4, 5	1, 2, 3
III	1	All	-	1, 2
IV	1	All	-	1, 2

Before each shot, the armor panel or plate shall be positioned as described in section 7.7.2.

After each shot: The bullet velocity shall be recorded. The actual location of the shot shall be examined to confirm that it complies with shot location requirements. The armor panel or plate and the backing material shall be examined to determine if a perforation occurred. All of the preceding observations shall be considered to determine if the shot was a fair hit, as defined in section 7.6.

If impact is not a fair hit: If an impact is determined to not be a fair hit, a second attempt shall be made immediately to obtain a fair hit in the same general location of the preceding unfair hit. No more than two attempts shall be made to obtain a fair hit at any general shot location. The maximum number of shots on any armor panel or plate shall not exceed the number specified in table 5 or table 6, as appropriate.

If more than two attempts are required to obtain a fair hit for any shot location, the test series on that panel shall be considered noncompliant. If the maximum number of allowable shots on the given armor panel/plate is exceeded, the test series on that panel/plate shall be considered noncompliant. Testing shall be repeated for any panel test series considered noncompliant. All test results, including the noncompliant test series, shall be reported.

If impact is a fair hit: If the impact did not produce a perforation and the shot was a fair hit, the BFS depression depth for the locations shown in table 7 shall be measured and recorded. If necessary, the backing material shall be struck (as described in section 4.2.5.4) or manipulated to produce an acceptable surface for remounting the armor.

7.8.5.1 BFS Measurement

The BFS shall be measured using a device capable of 1 mm (0.04 in) or better accuracy. The BFS shall be recorded. All measurements necessary to determine the BFS for a shot shall use a common reference surface. When the measured BFS exceeds 40 mm (1.58 in), the measurement shall be verified by a second measurement. Additionally, when the measurements are not fully automated and human interaction is required with the measuring device or backing material, a different individual shall make the second measurement.

7.8.6 P-BFS Test for Special Type Armor

The materials of construction and the construction details of special type armor shall be examined to determine the appropriate test methodology.

7.8.7 P-BFS Test for Accessory Ballistic Panels

7.8.7.1 Groin and Coccyx Protectors

Each protector shall be tested in its external carrier. Each protector shall be impacted with at least three evenly spaced fair hits at an angle of incidence of 0° (24 shots total). A single BFS depth measurement shall be taken on the first fair hit impact for each panel, for a total of eight BFS measurements.

7.8.7.2 Removable Side Protection

Removable side protection panels shall be tested in the primary armor's external carriers as appropriate for each armor model designation.

The samples shall be tested in accordance with the requirements of sections 7.6 and 7.7, placing as many fair hit impacts as possible, but not more than six shots, on each test sample to achieve a total of 24 fair hits per test caliber at the prescribed angles of incidence, per table 4. A single BFS depth measurement shall be taken on the first fair hit impact for each panel.

7.8.8 P-BFS Performance Requirements

Each test panel must withstand the appropriate number of fair hits and may not experience any perforations. Any complete perforation by a fair hit constitutes a failure. Fair hit is as defined in section 7.6.

New Samples: Each tested size of an armor model shall either have no BFS depth measurements that exceed 44 mm (1.73 in), or for each threat round an estimated probability of a single BFS depth measurement exceeding 44 mm (1.73 in) of less than 20 % with a confidence of 95 %.

The armor model shall be deemed to meet these requirements if no BFS depth measurement due to a fair hit exceeds 50 mm (1.97 in), and either:

(a) All BFS depth measurements due to fair hits are 44 mm (1.73 in) or less.

Or:

(b) The one-sided tolerance interval for a normal distribution indicates that there is 95 % probability that 80 % of the test BFS measurementsfor armor samples of that particular model, size, condition, and test threat will be 44 mm (1.73 in) or less. In this case the average, $\overline{Y}$, and sample standard deviation, s, of all recorded BFS measurements for armor samples of that particular model, size, condition, and test threat shall be calculated, and:

$$\overline{Y} + k_1 s \leq 44\,\text{mm}$$

For the case where there are 12 BFS measurements, $k_1 = 1.568$. For other quantities of BFS measurements, or for more details, see appendix D or Reference [9], section 7.2.6.3 (*http://www.itl.nist.gov/div898/handbook/prc/section2/prc263.htm*).

Conditioned Samples: For flexible armors that have been subjected to the conditioning protocol described in section 5, the BFS depth shall be recorded. However, an excessive BFS measurements will not constitute a failure.

Table 8. Baseline ballistic limit determination test summary

Armor Samples Required	Test Threat	Ballistic Panels Required	Minimum Shots Required	Minimum Penetration Results*
Type IIA through IIIA Approximately 5 armors per caliber.	Test Round 1	10	120	At least 60 stops. At least 30 perforations
	Test Round 2	10	120	At least 60 stops. At least 30 perforations
Type III 4 Armors	7.62 mm M80 FMJ	4	24	6 perforations, 12 stops, 6 either, velocity range of 27 m/s (90 ft/s)
Type IV 2 to 12 Armors	.30 caliber M2 AP	2 – 12**	12	3 perforations, 6 stops, 3 either, velocity range of 27 m/s (90 ft/s)

* When the armor's ballistic limit is sufficiently high that achieving the velocity necessary to perforate the armor is difficult or impossible, the test laboratory shall document that this situation has occurred. In such cases, the test will be considered acceptable even if the minimum number of perforations is not achieved.
** Quantity determined by section 4.1.2.2 and panel, plate, or insert size and ability to withstand multiple impacts.

7.9 Ballistic Limit (BL) Determination Test

The appropriate number of armor samples, as defined in section 4.1, shall be subjected to BL tests. The armor's ballistic performance will be estimated from the results. Each ballistic panel or plate shall be tested in accordance with the following procedures.

7.9.1 Requirements for Number of BL Tests and Test Samples

A complete BL test for an armor type consists of successful individual BL tests being conducted on the required number of armor samples specified in table 8. The armor panels or plates that comprise a sample shall each be subjected to a BL test using all threat rounds for which the armor will be certified (see table 4).

(a) <u>Flexible Vests and Jackets</u>: BL testing shall be conducted on complete armor samples (e.g., ballistic fabric panels, covers, carriers, and strapping). Removable trauma inserts/packs shall not be included as part of the complete armor sample used for BL determination.

(b) <u>Hard Plates</u>: Testing shall be conducted on complete armor samples, except when the armor's Type III or IV protection is provided entirely by rigid panels, plates, or inserts. In those instances, only the rigid panels, plates, or inserts shall be tested for baseline BL. When plate inserts are tested in conjunction with a flexible vest or jacket, the flexible component shall have previously demonstrated its full compliance with this standard at its appropriate level.

7.9.2 Test Procedure Requirements

The angles of incidence for all shots shall be 0°. All samples shall be tested dry.

All BL testing shall follow the basic procedures of sections 5.3.3 and 5.3.5 of MIL–STD–662F; however, the specific test parameters in table 9 shall be adopted. For each panel, shooting shall continue until either 12 shots or the maximum number of shots allowed on the panel has been reached. For armors that are limited to less than 12 shots, the firing sequence shall be continued on additional panels until 12 shots are reached. After each 12 shot sequence, a new sequence shall be performed until the total required number of shots as listed in table 8 has been reached.

Table 9. Test parameters and requirements for ballistic limit test

Parameter Description	Value	Section Reference in MIL–STD–662F
Velocity of first shot	The reference velocity for the armor type and caliber (table 4).	5.3.3
Velocity step until first reversal.	– 30.5 m/s (– 100 ft/s) if first shot was a perforation.	5.3.5
	+ 30.5 m/s (+ 100 ft/s) if first shot was a stop.	exception to 5.3.5
Velocity step until a second reversal.	± 22.9 m/s (± 75 ft/s), depending on result of previous shot	5.3.5
Velocity step after second reversal.	± 15.2 m/s (50 ft/s)), depending on result of previous shot	5.3.5

When the armor's ballistic limit is sufficiently high that achieving the velocity necessary to perforate the armor is difficult or impossible, the test laboratory shall document this situation has occurred. In such cases, the test will be considered acceptable even if the minimum number of perforations is not achieved.

7.9.3 Backing Material Conditioning

The backing material fixtures shall be prepared and conditioned to the same temperatures as those used to conduct the P-BFS tests for that armor model. Drop test validation shall be performed before each 12 shot series. However, post-test validation is not necessary. The backing material temperature shall be recorded before and after tests on a single armor panel/plate.

7.9.4 Data Set Tabulation

All shots fired for a BL test shall be reported. The shot information shall be reported in the order fired and shall include, at a minimum, shot number, velocity desired, actual velocity, and shot outcome.

7.9.5 Ballistic Limit Performance Requirements

All Armors: No perforations shall occur at or below the corresponding maxium P-BFS fair hit velocity, which is equal to the P-BFS reference velocity plus 9.1 m/s (30 ft/s).

New Armors: For new condition armors the BL test data shall be analyzed as described in appendix E, and the estimated probability of complete perforation at the corresponding P-BFS reference velocity must be less than 5 %. In other words, $\hat{V}_{05,new} \geq V_{ref,new}$.

REFERENCES

The following references form a basis for and support the procedures described in this standard:

[1] AATCC Recommended Washers and Dryers list, available from http://www.aatcc.org/Technical/WashDry.htm

[2] American National Standards Institute. SAAMI Z299.1–1992, *Voluntary Industry Standards for Pressure & Velocity of Rimfire Sporting Ammunition for the Use of Commercial Manufacturers*, 1992.

[3] American National Standards Institute. SAAMI Z299.3–1993, *Voluntary Industry Standards for Pressure and Velocity of Centerfire Pistol and Revolver Ammunition for the Use of Commercial Manufacturers*, 1993.

[4] American National Standards Institute. SAAMI Z299.4–1992, *Voluntary Industry Standards for Pressure & Velocity of Centerfire Rifle Sporting Ammunition for the Use of Commercial Manufacturers*, 1992.

[5] Department of Defense. MIL-STD-662F, *Department of Defense Test Method Standard, V50 Ballistic Test for Armor*, 1997 or latest version.

[6] Department of Defense. MIL-STD-810F, *Department of Defense Test Method Standard for Environmental Engineering Considerations and Laboratory Tests*, 2000 or latest version.

[7] National Institute of Justice. NIJ Standard–0101.04, *Ballistic Resistance of Police Body Armor*, 2000.

[8] *NIST/SEMATECH e-Handbook of Statistical Methods*, http://www.itl.nist.gov/div898/handbook/, May 2008.

[9] U.S. Army Test and Evaluation Command. TOP 2–2–710, *Test Operations Procedure, Ballistic Tests of Armor Materials*, 1984 or latest version.

APPENDIX A – ACCEPTABLE BULLETS FOR HANDLOADING

Table 10 below lists the acceptable bullets for testing. For the purpose of inter-laboratory consistency, for threat levels IIA, II, and IIIA, only those model numbers specified in table 10 shall be acceptable.[1] For threat levels III and IV, bullets from alternate manufacturers may be used if they meet the specified weight, geometry, jacket, and core material requirements.

All jacket materials shall be of copper or copper alloy (approximately 90 % copper and 10 % zinc), with the exception of Type III, which shall be steel.

Table 10. Acceptable bullets

Threat Level	Caliber	Bullet Weight	BULLET DESCRIPTION	Nominal Bullet Diameter	Acceptable Manufacturer	Bullet Model Number
IIA	9 mm Luger	8.0 g (124 gr)	FMJ RN	9 mm (.355 in)	Remington	23558
IIA	.40 S&W	11.7 g (180 gr)	FMJ	10 mm (.400 in)	Remington	23686
II	9 mm Luger	8.0 g (124 gr)	FMJ RN	9 mm (.355 in)	Remington	23558
II	.357 Mag	10.2 g (158 gr)	JSP	9.1 mm (.357 in)	Remington	22847
IIIA	.357 SIG	8.1 g (125 gr)	TMJ	9.0 mm (.355 in)	Speer	4362
IIIA	.44 Mag	15.6 g (240 gr)	JHP	10.9 mm (.429 in)	Speer	4453 or 4736**
III	7.62 mm NATO	9.6 g (147 gr)	FMJ – SPIRE PT BT*	7.62 mm (.308 in)	***	***
IV	30.06 M2 AP	10.8 g (166 gr)	FMJ – SPIRE PT AP	7.62 mm (.308 in)	May be obtained from U.S. Military M2 AP ammunition	

*   Verify that jacket is ferrous (use of a magnet is acceptable).
** Note: These two models are the same bullet but sold in different quantities.
*** Bullet may be obtained from U.S. military/NATO M80 ammunition, or from other manufacturers meeting the specifications for the projectile in the M80 cartridge.

---

[1] The use of brand names and model numbers in this standard does not constitute endorsement by the U.S. Department of Justice; Office of Justice Programs; National Institute of Justice; U.S. Department of Commerce; National Institute of Standards and Technology; Office of Law Enforcement Standards; or any other agency of the United States Federal Government, nor does it imply that the product is best suited for other applications.

APPENDIX B – COMMON SPECIAL TYPE THREATS

Table 11 lists some common threat rounds that may be problematic for some armors. These threats, with their accompanying test velocities, may be used in the special type testing to validate that a particular armor model may stop these rounds in addition to the standard threats.

The threat rounds listed in the table are not the only rounds that can be tested in the special type tests. A purchaser or manufacturer may specify any round for testing with the special type tests. These are included in the standard only to simplify the procedure of specifying common special threat rounds.

Table 11. Special type threats of particular concern to law enforcement

Manufacturer	Model	Caliber	Weight	Description	Diameter	Nominal (Factory) Velocity (ft/s)	Recommended Conditioned Test Velocity (ft/s)	Recommended New Armor Test Velocity (ft/s)
Federal	Tactical Bonded	9mm Luger	135 gr	Tactical HP	9 mm (0.355 in)	1060	1120	1150
Federal	Tactical Bonded	.357 SIG	125 gr	Tactical HP	9 mm (.355 in)	1350	1410	1440
Federal	Tactical Bonded	.40 S&W	165 gr	Tactical HP	10 mm (.400 in)	1050	1110	1140
Federal	Tactical Bonded	.40 S&W	180 gr	Tactical HP	10 mm (.400 in)	1000	1060	1090
Federal	Tactical Bonded	.45 ACP	230 gr	Tactical HP	11.5mm (.451 in)	950	1010	1040
Speer	Gold Dot	9 mm Luger	124 gr	GDHP	9 mm (0.355 in)	1220	1280	1310
Speer	Gold Dot	.357 SIG	125 gr	GDHP	9 mm (0.355 in)	1375	1435	1465
Speer	Gold Dot	.40 S&W	165 gr	GDHP	10 mm (.400 in)	1150	1210	1240
Speer	Gold Dot	.40 S&W	180 gr	GDHP	10 mm (.400 in)	1025	1085	1115
Speer	Gold Dot	.45 ACP	185 gr	GDHP	11.5mm (.451 in)	1050	1110	1140
Speer	Gold Dot	.45 ACP	230 gr	GDHP	11.5mm (.451 in)	890	950	980
Winchester	Ranger T-Series	9 mm Luger	127 gr	JHP	9 mm (0.355 in)	1250	1310	1340
Winchester	Ranger T-Series	.357 SIG	125 gr	JHP	9 mm (0.355 in)	1350	1410	1440
Winchester	Ranger T-Series	.40 S&W	165 gr	JHP	10 mm (.400 in)	1140	1200	1230
Winchester	Ranger T-Series	.40 S&W	180 gr	JHP	10 mm (.400 in)	990	1050	1080
Winchester	Ranger T-Series	.45 ACP	230 gr	JHP	11.5mm (.451 in)	990	1050	1080
FN	SS192	5.7 mm	28 gr	JHP	5.7 mm (.224 in)	2050	2110	2140
FN/Hornady	SS197SR	5.7 mm	40 gr	V-Max	5.7 mm (.224 in)	1700	1760	1790

APPENDIX C – ARMOR SIZING TEMPLATES

The following five templates are intended for use with most models of flexible armor in the form of concealable or tactical vests or jackets. The same templates may be used for male, female, or unisex armor designs. The five templates are:

[1] Template NIJ–C–1 (Smallest)
[2] Template NIJ–C–2 (Small)
[3] Template NIJ–C–3 (Medium)
[4] Template NIJ–C–4 (Large)
[5] Template NIJ–C–5 (Largest)

The dimensions of these templates are shown in drawings on the following five pages, and the maximum and minimum areas are shown in table 12 below. Table 13 and table 14 list the smallest and largest allowable production armors for each of the available template sizes.

Table 12. Surface areas of armor sizing templates

Template	Maximum Area (Largest Rear Panel)	Minimum Area (Smallest Front Panel)
NIJ–C–1	0.0939 m^2 (146 in^2)	0.0659 m^2 (102 in^2)
NIJ–C–2	0.1354 m^2 (210 in^2)	0.1020 m^2 (158 in^2)
NIJ–C–3	0.1835 m^2 (284 in^2)	0.1443 m^2 (224 in^2)
NIJ–C–4	0.2393 m^2 (371 in^2)	0.1945 m^2 (301 in^2)
NIJ–C–5	0.3022 m^2 (468 in^2)	0.2517 m^2 (390 in^2)

Table 13. Minimum allowable surface areas for production armor

If the smaller template tested is:	The minimum area of production armor shall be:
NIJ–C–1	No limit
NIJ–C–2	0.0980 m^2 (152 in^2)
NIJ–C–3	0.1399 m^2 (217 in^2)
NIJ–C–4	0.1890 m^2 (293 in^2)
NIJ–C–5	Not applicable

Table 14. Maximum allowable surface areas for production armor

If the larger template tested is:	The maximum area of production armor shall be:
NIJ–C–1	Not applicable
NIJ–C–2	0.1399 m^2 (217 in^2)
NIJ–C–3	0.1890 m^2 (293 in^2)
NIJ–C–4	0.2455 m^2 (381 in^2)
NIJ–C–5	No limit

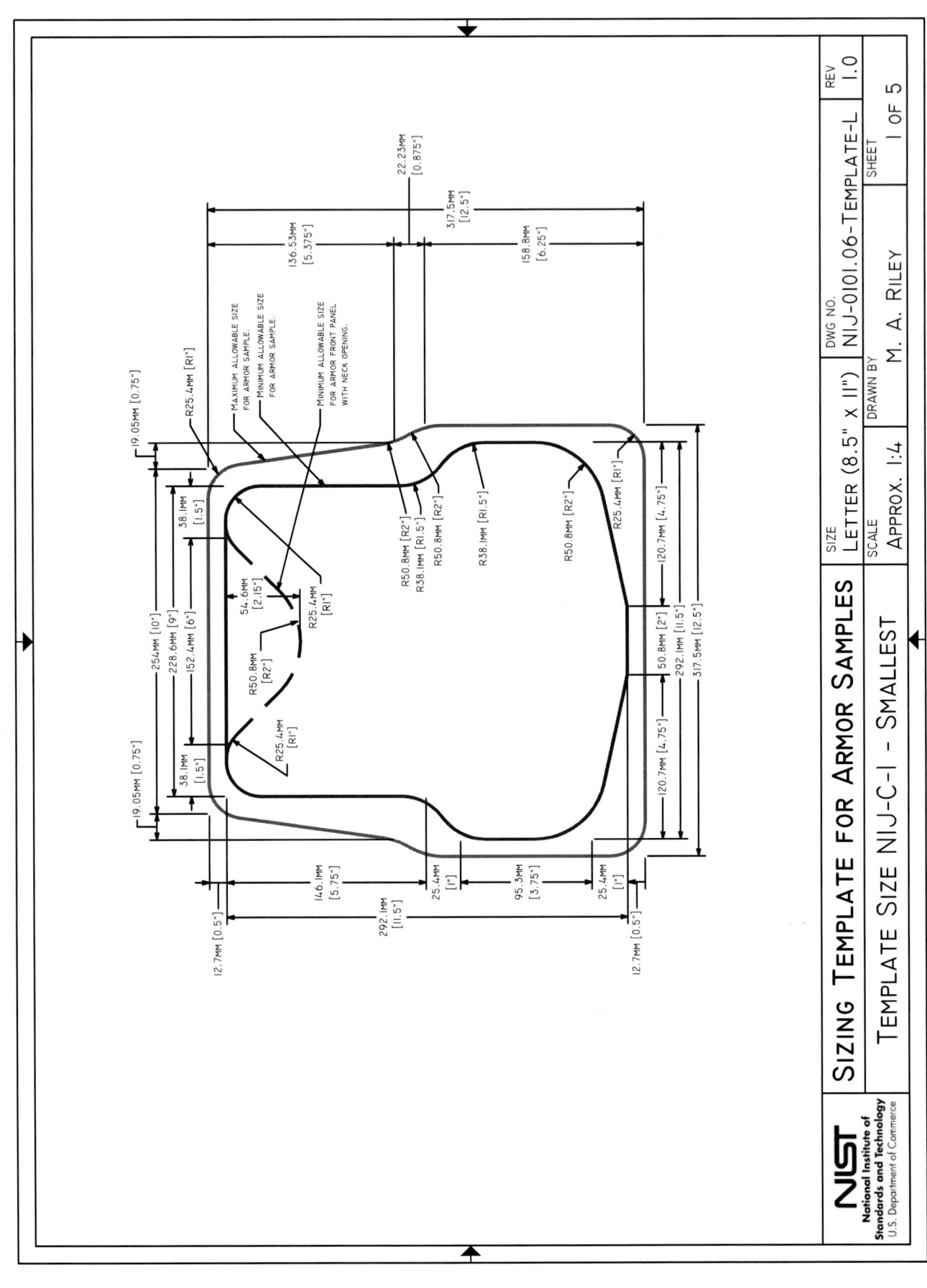

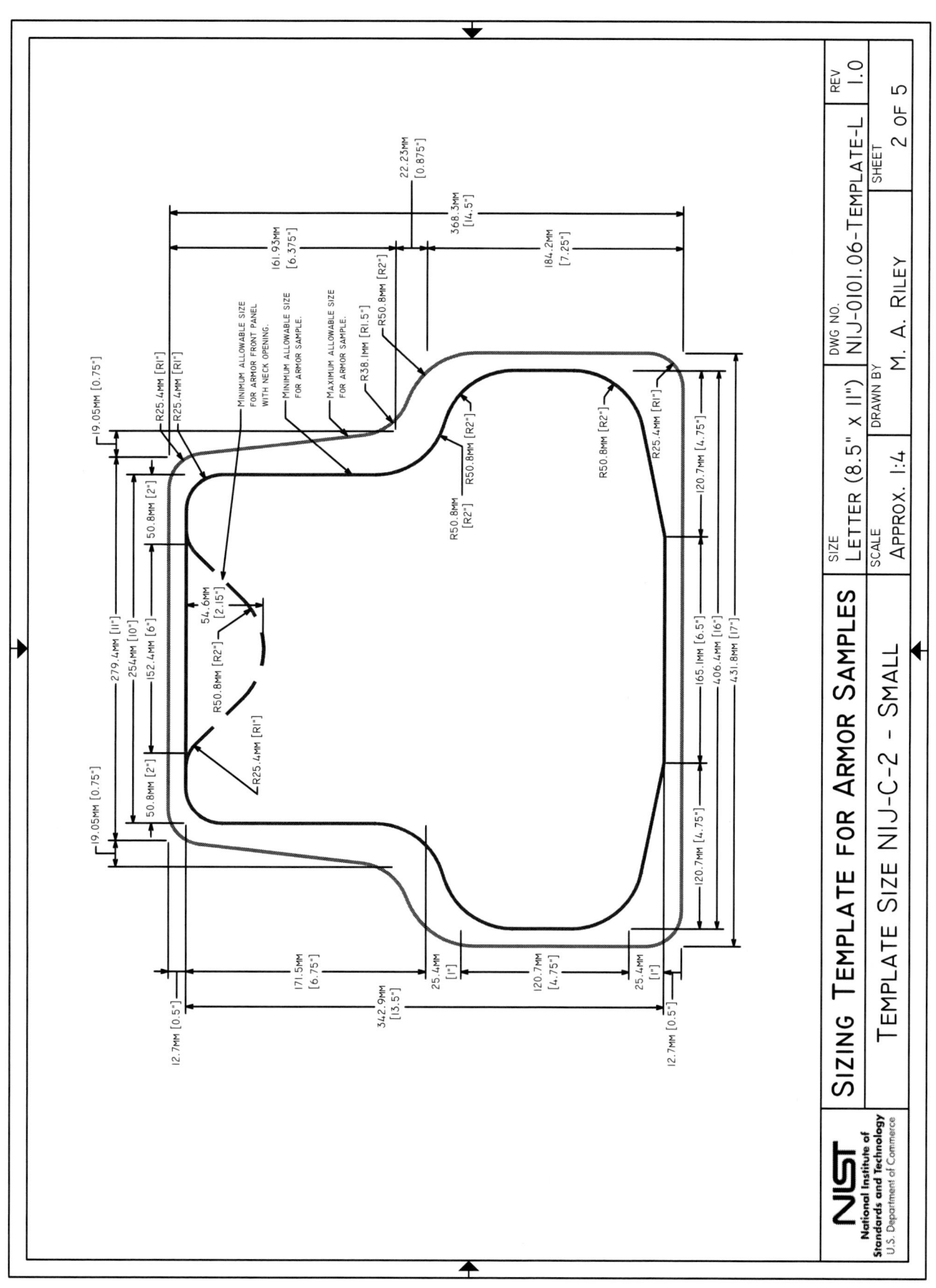

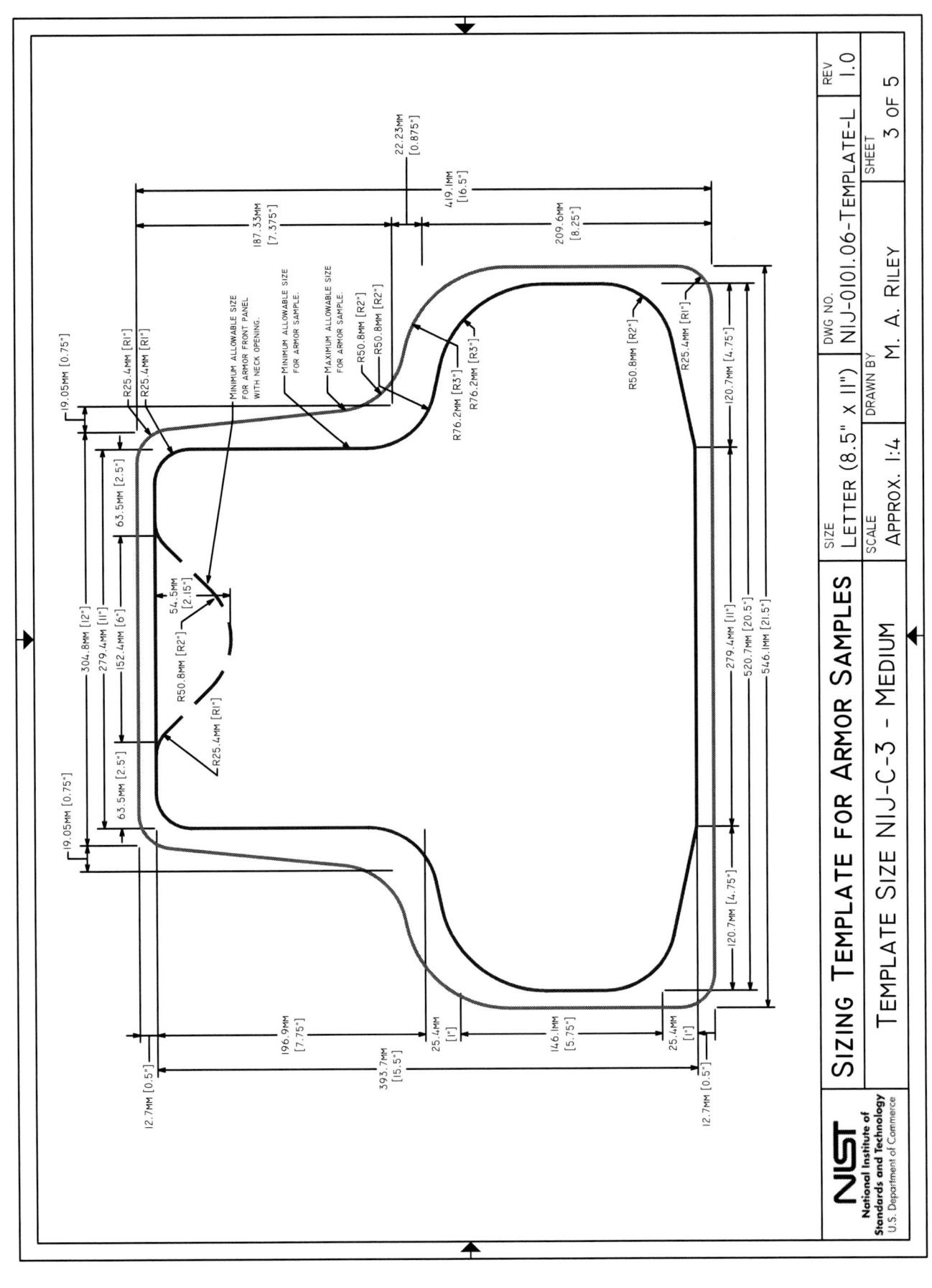

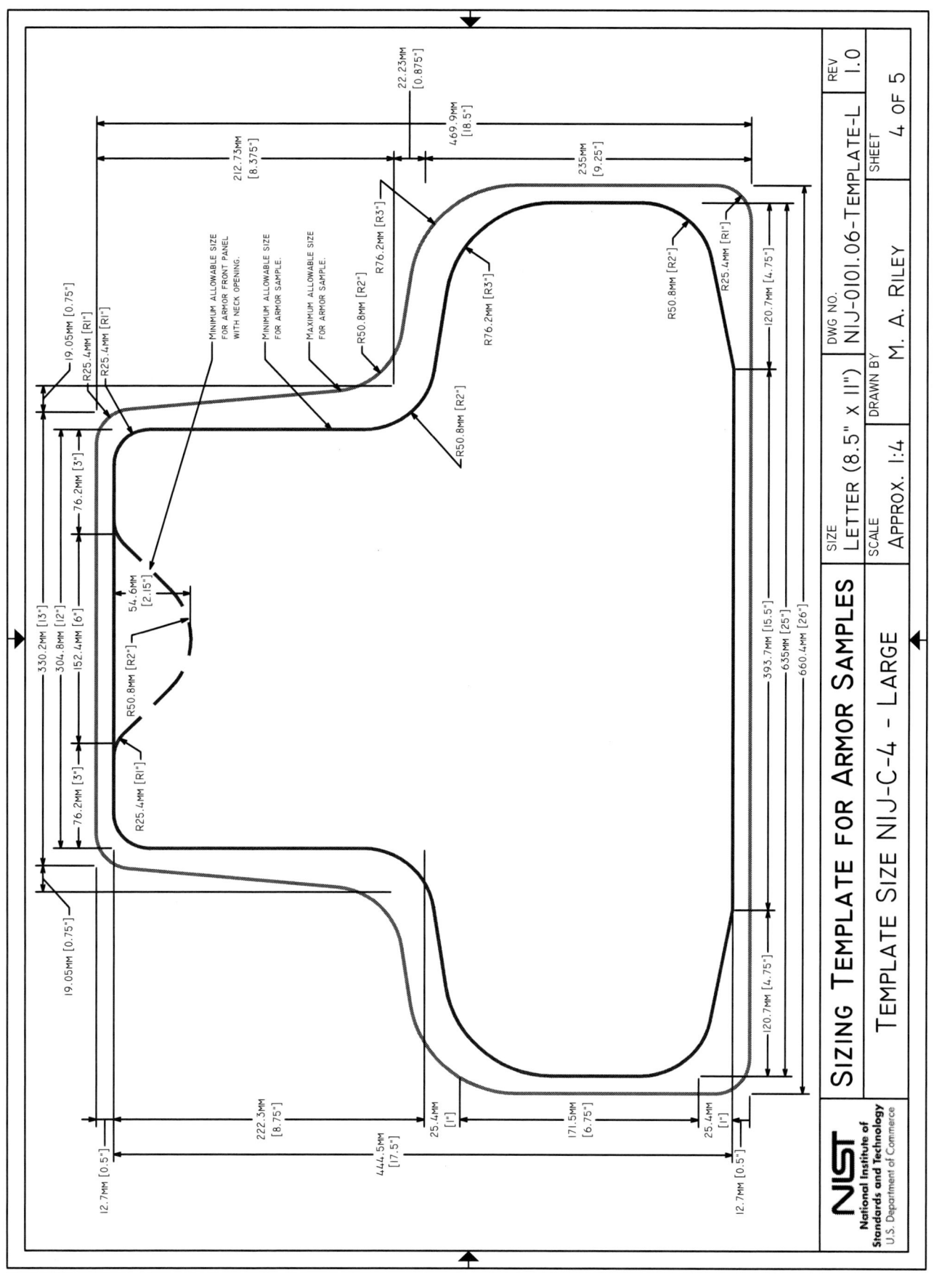

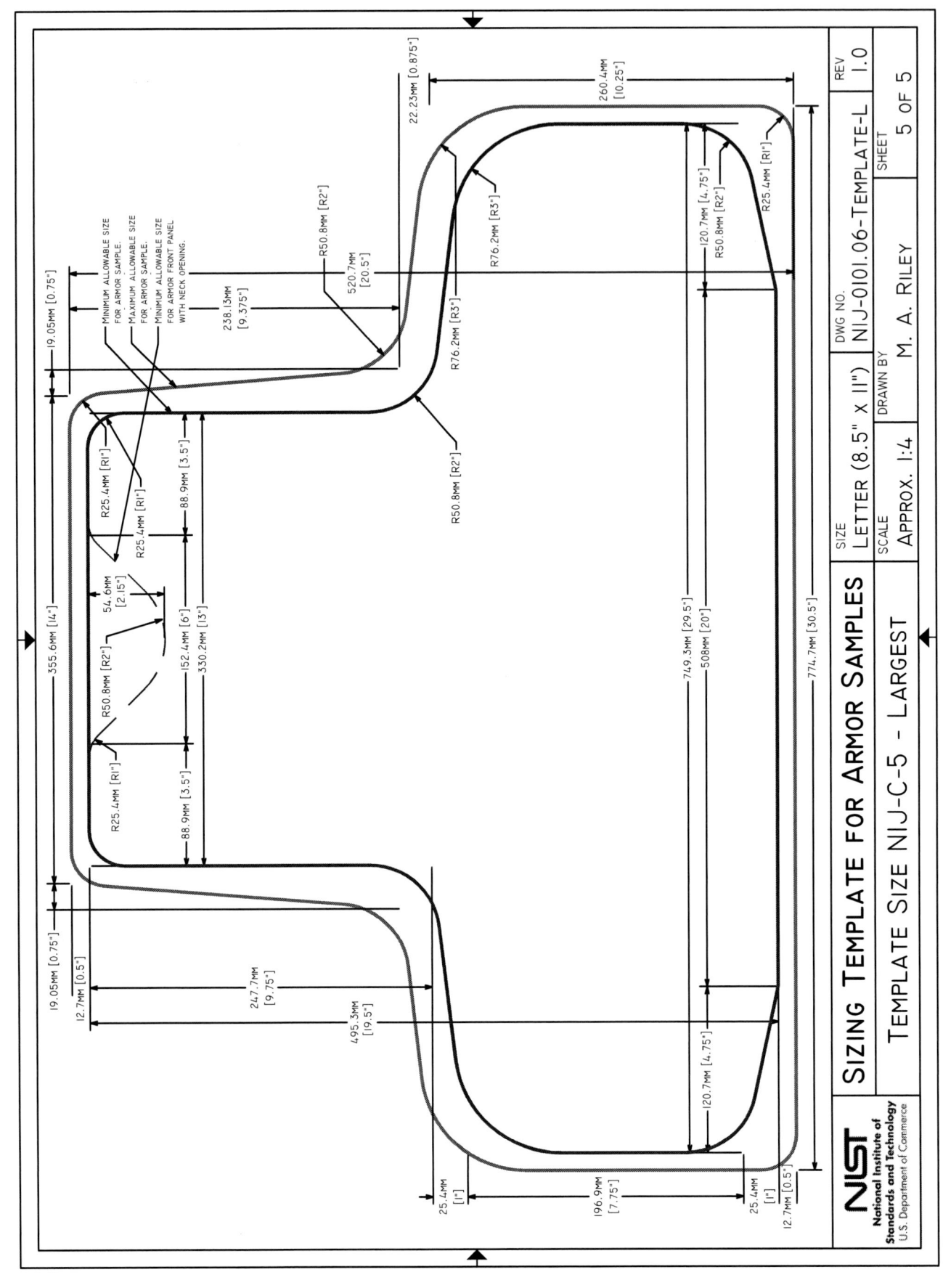

APPENDIX D – ANALYSIS OF BACKFACE SIGNATURE DATA

The measured backface signatures from a P-BFS test for new armor shall be analyzed to determine if the armor will provide adequate protection against behind armor blunt trauma. The requirements given in section 7.8.8 specify that either all measured BFS depths due to fair hits shall be 44 mm (1.73 in) or less, or if any BFS depth exceeds 44 mm (1.73 in) then there shall be 95 % confidence that 80 % of all BFS depths will be 44 mm (1.73 in) or less. In no case may a BFS depth exceed 50 mm (1.97 in).

The requirements of the second condition can be verified using a statistical tolerance limit (see section 8, Reference [8]). In this case, we expect a stated portion of the entire population of all BFS measurements to lie at or below the statistical upper tolerance limit. To achieve this, the population of BFS measurements is assumed to be normally distributed, and the upper tolerance limit, Y_U, must be less than or equal to 44 mm (1.73 in). The upper tolerance limit is defined as:

$$Y_U = \overline{Y} + k_1 s$$

Here, $\overline{Y}$ is the average of all BFS measurements for armor samples of that particular model, size, condition, and test threat; s is the sample standard deviation of the same set of BFS measurements; and k_1 is a factor that must be determined such that the interval covers the appropriate proportion, p, with a confidence of γ.

The average, $\overline{Y}$, is simply calculated as:

$$\overline{Y} = \frac{1}{N}\sum_{i=1}^{N} Y_i$$

Here, N is the number of BFS measurements, and Y_i are the individual BFS measurements. The sample standard deviation, s, is then calculated as:

$$s = \sqrt{\frac{1}{N-1}\sum_{i=i}^{N}(Y_i - \overline{Y})^2}$$

The approximate k factor, k_1, for a one-sided tolerance interval can now be calculated as (see section 8, Reference [8]):

$$k_1 = \frac{z_{1-p} + \sqrt{z_{1-p}^2 - ab}}{a}$$

Here, z_{1-p} is the normal distribution critical value that is exceeded with a probability of $1-p$. The factors a and b are defined as:

$$a = 1 - \frac{z_{1-\gamma}^2}{2(N-1)}; \qquad b = z_{1-p}^2 - \frac{z_{1-\gamma}^2}{N}$$

Here, $z_{1-\gamma}$ is the normal distribution critical value that is exceeded with a probability of $1-\gamma$.

For the analysis of BFS measurements according to the requirements of this standard, the probability that no BFS measurement exceeds 44 mm (1.73 in) must be at least 80 %, so $p = 0.80$, and the required confidence is 95 %, so $\gamma = 0.95$. The critical values for the normal distribution can be calculated or obtained from tables in many statistical texts. For this case they are:

$$z_{1-\gamma} = z_{0.05} = 1.645; \qquad z_{1-p} = z_{0.20} = 0.842$$

Using these results, the factors a and b can be calculated for a given number, N, of BFS measurements. For $N = 12$, factors a and b are:

$$a = 1 - \frac{1.645^2}{2(12-1)} = 0.877; \qquad b = 0.842^2 - \frac{1.645^2}{12} = 0.483$$

Then, the k factor, k_1, is:

$$k_1 = \frac{0.842 + \sqrt{0.842^2 - (0.877)(0.483)}}{0.877} = 1.568$$

Other k factors, for quantities of BFS measurements that will be typical of tests performed to this standard, are listed in table 15.

The allowable excessive BFS probability, 20 %, may appear to be high; however, this value is intended to account for both the variation in the armor's performance, which should be small, and the variation in the BFS measurement due to the backing material and the backing material preparation. While careful treatment and preparation of the backing material by the test laboratory can minimize the variation due to the backing material, there will always be some inherent variation introduced into the test results by the backing material. The required probability is chosen to reduce that possibility that an acceptable armor design will fail the P-BFS test due to reasonable variation in the backing material.

Table 15. k factors for 80 % probability with 95 % confidence

Number of BFS Measurements, N	k factor, k_1
6	2.143
7	1.961
8	1.837
9	1.745
10	1.673
11	1.616
12	1.568

APPENDIX E – ANALYSIS OF BALLISTIC LIMIT DATA

Once the ballistic limit testing has been completed, the test results should be analyzed for each test threat by performing a regression to estimate what the armor's performance will be over a range of velocities. In particular, the analysis should attempt to estimate the velocity where the probability of perforation becomes reasonably small. In general a logistic regression can be used for this purpose; however, other probability distributions and regression methods may be used when one can be shown to better estimate the performance of a particular armor model.

The logistic regression may be performed on the data using the method of maximum likelihood to estimate the logistic parameters $\hat{\beta}_0$ and $\hat{\beta}_1$, which are the estimated *logistic constant* and the estimated *velocity coefficient*, respectively. These parameters define the shape of the S-shaped logistic curve, which is defined as:

$$\pi(v) = \frac{e^{\beta_0 + v\beta_1}}{1 + e^{\beta_0 + v\beta_1}}$$

Here $\pi(v)$ is the probability of a complete perforation occurring at velocity, v. From the estimated logistic parameters, the ballistic limit can be determined as:

$$\hat{V}_{50} = \frac{-\hat{\beta}_0}{\hat{\beta}_1}$$

In addition, the velocity at which the probability of a complete perforation is $x\%$, $\hat{V}_x$, can be determined as:

$$\hat{V}_x = \frac{\ln\left(\frac{x}{1-x}\right) - \hat{\beta}_0}{\hat{\beta}_1}$$

The estimated logistic parameters for a conditioned armor and its ballistic limit can be determined in the same method; however, care should be exercised when the analysis is performed on a relatively small data set, as the reliability of the estimated perforation probability will be poor for small data sets.

APPENDIX F – EXPLANATORY MATERIALS

The following information is provided as explanation for the sections indicated.

Explanatory Material for Section 2 NIJ Body Armor Classification

The ballistic threat posed by a bullet depends on its composition, shape, caliber, mass, angle of incidence, and impact velocity, among other things. Because of the wide variety of bullets and cartridges available in a given caliber and the existence of handloaded ammunition, armors that will defeat a standard test round may not defeat other threats of the same caliber. An armor that defeats a given lead bullet may not resist perforation by other bullets of the same caliber having different construction or configuration. The test ammunitions specified in this standard represent higher velocity versions of threats that law enforcement officers may face in the United States, but which also are among the more difficult threats to safely stop. By testing armors against these threats, the armor will generally be able to stop a wide variety of similar and lesser threats.

As of the publication of this standard, ballistic resistant body armor suitable for full-time wear throughout an entire shift of duty is available in classification Types IIA, II, and IIIA, which provide increasing levels of protection from handgun threats. Type IIA body armor will provide minimal protection against smaller caliber handgun threats. Type II body armor will provide protection against many handgun threats, including many common, smaller caliber pistols with standard pressure ammunition, and against many revolvers. Type IIIA body armor provides a higher level of protection, and will generally protect against most pistol calibers, including many law enforcement ammunitions, and against many higher powered revolvers.

Types III and IV armor, which protect against rifle rounds, are generally used only in tactical situations or when the threat warrants such protection.

Type I body armor, which was first issued during the NIJ demonstration project in 1975, has been removed from this test standard due to the increasing prevalence of higher powered threats and the increased power of most law enforcement duty weapons. While it is not yet necessary to remove Type IIA armor from service, agencies that are using this level of protection are advised to review the threats they face and to consider upgrading to a higher level of protection when their current armor reaches the end of its service life.

Explanatory Material for Section 3 Definitions

The definition of perforation was added to improve the clarity of the document and make the terminology consistent with international standards. *Perforation* replaces *complete penetration*. Although the terms *partial penetration* and *complete penetration* are no longer used in this standard, they may still be used for compatibility with military standards.

Explanatory Material for Section 4 Sample Requirements and Laboratory Configuration

4.1 Test Samples

Ballistic limit testing is not required on smaller sized armor panels for the following reasons. Past research has shown that the size of the armor generally has only a small impact on the penetrability of the armor and that the larger armors tend to be slightly easier to perforate. Therefore, the larger armors are expected to perform slightly worse than the smaller armors. Additionally, there is not enough area on the smaller armors for good ballistic limit testing, and many more panels would be required for the smaller armors.

4.1.5 Labeling

The requirement for English on the label has been removed for the sake of allowing this standard to be more widely used. The requirement for use of a particular language for the label markings should be inserted into the appropriate conformity program or procurement requirements.

4.2.2.1 Handloads

Test laboratories are advised to consider factors that might produce variable ballistic results (day-to-day variations, powder lot, bullet lot, etc.), and to confirm that required velocities will be achieved before testing armor. For ballistic limit testing, sufficiently accurate records should be kept for each test threat such that the load required for particular velocity can be estimated with reasonable accuracy.

4.2.2.3 Test Barrel Fixtures

Test laboratories are advised to verify that correct velocities will be achieved, ensure that the test barrel has achieved thermal stability, and use an appropriate aiming system to ensure proper placement of the test bullet.

4.2.4 Armor Submersion Equipment

No formal test method is provided for measuring impurities in the water used for armor submersion. "Visible impurities" refers to seeing any debris, material, or particles floating in the water or seeing discoloration in the water. Manufacturers may request fresh water for armor submersion of their products.

Explanatory Material for Section 5 Flexible Armor Conditioning Protocol

This method may not reproduce all of the humidity effects associated with the natural environment, such as long-term effects of exposure to high humidity or to low humidity situations. This method does not attempt to duplicate the complex temperature/humidity environment but provides a generally stressful situation that is intended to reveal potential problem areas in the armor.

Accordingly, this procedure does not reproduce naturally occurring or service-induced temperature-humidity time histories, nor is it intended to produce humidity effects that have been preceded by solar effects. It may induce problems that are indicative of long-term effects.

Explanatory Material for Section 6 Hard Armor Conditioning Protocol

This method may not reproduce all of the humidity effects associated with the natural environment, such as long-term effects of exposure to high humidity or to low humidity situations. This method does not attempt to duplicate the complex temperature/humidity environment but, rather, provides a generally stressful situation that is intended to reveal potential problem areas in the armor. This test method will not predict the service life of the armor. Accordingly, this procedure does not reproduce naturally occurring or service-induced temperature-humidity time histories, nor is it intended to produce humidity effects that have been preceded by solar effects. It may induce problems that are indicative of long-term effects. This method does not simulate an exact period of time in the field, nor is it intended as an absolute predictor of actual armor service life.

There is no armor conditioning requirement for the in conjunction flexible armor portions. The plate inserts are required to undergo conditioning, and the flexible armor shall be listed on the NIJ Compliant Products List, indicating that they have already passed both conditioning and ballistic testing.

Explanatory Material for Section 7 Ballistic Test Methods

7.3 Workmanship Examination

Photographs of all samples are not necessary; however, pretest and post-test photographs of all samples, along with photographs of any deficiencies, failures, or unusual results can provide useful information to both the armor manufacturer and the conformity assessment body, and are, therefore, preferred. Such photographic documentation may be required by either the manufacturer or the CTP.

7.6.1 Minimum Shot-to-Edge Distance

The minimum shot-to-edge distance for many test threats is now 51 mm (2.0 in), as opposed to 76 mm (3.0 in) in NIJ Standard–0101.03 and NIJ Standard–0101.04. This returns the limit to distance it was at in earlier versions of the standard.

For the heavier threat rounds used against the Type IIA, II, and IIIA armors, the minimum shot-to-edge distance remains at 76 mm (3.0 in).

The minimum shot-to-edge distance may be further reduced for a particular model at the request of the manufacturer.

7.8 Ballistic Perforation and Backface Signature Testing

Neither the clay backing material nor the backface signature depth measurement reflects characteristics of the human torso or its response to ballistic impact. The clay backing material provides a medium for making BFS measurements.

Explanatory Material for Section 7.9.5

While not a requirement of the current standard, the ballistic limit results may be used to estimate whether a conditioned or a field return has degraded to the point where its performance may be questionable.

From the analysis described above, an acceptable degradation margin for aged armors, V_{margin}, can be defined as:

$$V_{margin} = \min \left\{ \begin{array}{c} \hat{V}_{LP,new} \\ \hat{V}(\hat{\pi}_{95\%,up} \leq 0.05) \end{array} \right\} - V_{ref}$$

This degradation margin is based on the assumption that while the armor's performance will have declined, the velocity coefficient of the performance curve will have remained nearly the same. Once this margin has been estimated, and assuming that while the armor's performance will have declined, the velocity coefficient of the performance curve will have remained nearly the same, a minimum allowable aged armor ballistic limit may be established. For aged armors the ballistic limit should not have degraded more than the degradation margin. Due to the limited amount of data available to determine the aged armor ballistic limit, some additional reduction might be allowed to account for the variation in the aged armor ballistic limit estimate. This leads to a minimal aged armor ballistic limit that can be defined as:

$$\hat{V}_{50,aged} \geq \hat{V}_{50,new} - (V_{margin} + 15\text{m/s})$$

Or:

$$\hat{V}_{50,aged} \geq \hat{V}_{50,new} - (V_{margin} + 49\text{ft/s})$$

Care should be taken when analyzing response data from aged, and particularly field-returned, armors. Ballistic limit test data from armors that have not been aged in the same way should generally not be lumped together, and the results from any single specimen may not be typical of an armor model. However, if estimated ballistic limits from more than a small percentage of aged armors are either close to or less than the established minimum, there may be reason to be concerned with the armor's long term-performance.

INDEX

A

accuracy, 25, 27, 33, 48, 72
agents, 8
air, 5, 7, 9, 27, 29, 33
alcohol, 18
alternative, 41
ambient air, 7
amplitude, xi
analog, 29, 35
appendix, 4, 11, 22, 23, 50, 52
applied research, ix
arithmetic, 22, 24
Army, 53
assessment, 73
attachment, 20

B

benchmarks, ix
bust, 8, 44

C

caliber, 4, 22, 49, 71
calibration, 27, 33
carrier, 5, 7, 18, 39, 40, 49
chemicals, 28, 35
chlorine, 28, 35
classes, 3
classification, 3, 71
classified, 3
clay, 6, 9, 23, 24, 36, 41, 43, 74
closure, 44
coatings, 7
coccyx, 5, 17
commercial, ix, 7, 28, 35
community, 1
compatibility, 7, 9, 71
compliance, ix, 11, 12, 13, 17, 30, 39, 40, 43, 45, 51
components, 5, 24, 39, 40
composition, 71
concrete, 34
condensation, 7, 29
conditioning, 5, 12, 19, 20, 24, 27, 29, 31, 33, 36, 39, 45, 50, 73
confidence, 49, 65, 66, 67
configuration, 8, 21, 40, 41, 43, 71
conformity, 23, 72, 73
construction, 17, 20, 39, 41, 44, 49, 71
contaminants, 28, 35
control, 27, 33
controlled, 27, 30, 36
copper, 7, 8, 9, 55
corrosive, 28, 35
cotton, 18, 20
coverage, 7
covering, 9
criminal justice, v, ix, 1
critical value, 66

D

data set, 69
decibel, xi
deficiency, 39
definition, 71
degradation, 74
degree, xi
demineralized, 23
denatured, 18
density, 20
Department of Commerce, 23, 55
Department of Defense, 53
Department of Justice, ix, 23, 55

depression, 23, 41, 43, 48
deviation, 10
dew, 7
discontinuity, 44
distillation, 28, 35
distilled water, 18
drop test, 23, 24, 25, 36
dry, 45, 51
durability, 18, 19, 33, 34, 38
duration, 28, 45

E

electromotive force, xi
electronic, 21
English, 72
environment, 27, 30, 33, 34, 36, 72, 73
equilibrium, 10
equipment, ix, 7, 20, 21, 23, 24, 27, 33
evidence, 17
exposure, 5, 30, 33, 34, 36, 72, 73

F

fabric, 10, 17, 51
failure, 29, 35, 49, 50
February, 38
fibers, 10
film, 10
fire, 18, 22
firearms, 22
flight, 5, 6, 10, 43
floating, 72
flow, 29, 31, 36
fresh water, 72

G

grain, xi
group size, 13
guidance, 44
guidelines, ix, 27, 33

H

hazards, 1
heat, 27, 33
height, 24, 36, 44
Homeland Security Act, ix
homogenous, 6
human, 49, 74

humidity, xi, 5, 7, 20, 27, 28, 29, 30, 33, 34, 35, 36, 72, 73

I

identification, 17, 18
impurities, 23, 28, 35, 72
incidence, 5, 21, 40, 43, 45, 48, 49, 51, 71
indication, 27, 33
infrared, xi
injection, 28, 35
instruments, 1, 18, 27, 33
interaction, 49
interactions, 6
international standards, 71
International Standards Organization (ISO), xi, 27, 33
interval, 49, 65

J

justice, ix

L

labeling, 17, 39
language, 72
laser, 24
law enforcement, ix, 58, 71
lead, 7, 8, 9, 71
likelihood, 69
limitations, 44
linear, 6
liquid phase, 10
location, 3, 18, 20, 33, 40, 43, 45, 48
long-term, 72, 73
lumen, xi

M

magnet, 55
maintenance, 28, 34
manufacturer, 10, 11, 12, 17, 18, 19, 36, 39, 40, 41, 45, 57, 73
material surface, 8, 25
measurement, 4, 6, 9, 22, 23, 25, 29, 35, 41, 46, 49, 66, 74
mechanical, 5, 27, 30, 33, 34, 38
media, 43
military, 7, 9, 71
modeling, 6, 23

models, 1, 13, 17, 18, 20, 43, 55, 59
modulation, xi
moisture, 27, 33

N

nanometer, xi
national, ix
National Institute of Standards and Technology (NIST), ix, 23, 53, 55
NATO, 55
natural environment, 72, 73
normal distribution, 49, 66

O

observations, 48
Office of Justice Programs, ix, 23, 55
oil, 6, 23
orientation, 17, 18, 35, 43
osmosis, 28, 35

P

parameter, 28, 34
particles, 72
penetrability, 72
perforation, 6, 7, 9, 39, 40, 43, 48, 49, 52, 69, 71
performance, ix, 1, 3, 8, 9, 20, 27, 33, 39, 46, 50, 66, 69, 74
periodic, 28, 29, 34
permit, 21
personal, 1, 7
photographs, 39, 73
planar, 41
plastic, 10
poor, 39, 69
population, 65
powder, 72
power, 71
preparation, 23, 24, 66
pressure, 7, 10, 71
probability, 49, 52, 66, 67, 69
procedures, v, 43, 50, 51, 53
production, 20, 59
program, ix, 72
property, 1
protection, 3, 4, 5, 7, 8, 9, 12, 17, 18, 49, 51, 65, 71
protocol, 12, 19, 20, 27, 29, 33, 35, 39, 45, 50

R

radio, xi
random, 12
range, 11, 21, 22, 28, 35, 40, 69
real time, 28, 34
recovery, 24
reduction, 74
regression method, 69
reliability, 69
repair, 24
representative samples, ix
research, 72
resistance, 1, 17, 39, 43
resistivity, 28, 35
returns, 29, 73
rotations, 28

S

safeguards, 1
safety, vi
sample, 5, 7, 10, 11, 12, 17, 19, 22, 30, 39, 43, 49, 51, 65
sensors, 27, 33, 34, 35
series, 24, 25, 39, 43, 44, 45, 48, 52
shape, 23, 69, 71
single test, 11
software, 35
solar, 73
S-shaped, 69
stability, 21
standard deviation, xi, 49, 65
standards, ix, xi, 1, 7, 9, 23, 53, 55, 71
steel, 4, 24, 55
storage, 27, 33, 34
supply, 41
surface area, 59
systems, 7, 27, 33

T

technological, ix
technology, ix, 28, 34
temperature, 5, 7, 9, 23, 24, 27, 28, 29, 30, 33, 35, 36, 52, 72, 73
test data, 52, 74
test procedure, 39
textile, 7
The Homeland Security Act, ix
thermal, 33, 34, 72

threat, 4, 8, 10, 11, 12, 17, 21, 25, 40, 43, 44, 49, 50, 51, 55, 57, 65, 69, 71, 72, 73
threats, 1, 3, 4, 7, 8, 9, 11, 40, 45, 57, 58, 71, 73
time, 6, 21, 24, 27, 29, 30, 33, 35, 45, 71, 73
tolerance, 21, 22, 23, 28, 29, 49, 65
trauma, 8, 20, 39, 40, 51, 65

U

U.S. military, 55
ultraviolet, xi
uncertainty, 22
uniform, 27, 33, 34
United States, 23, 55, 71

V

validation, 24, 52
vapor, 7, 10
variable, 72
variance, 17
variation, 66, 74

velocity, 3, 4, 6, 7, 8, 9, 22, 40, 46, 48, 50, 52, 69, 71, 72, 74
visible, 9, 18, 23
visual, 28, 30, 34, 36
voids, 23, 24, 41

W

warrants, 71
Washington, vi
water, 5, 7, 9, 23, 28, 29, 35, 45, 72
water supplies, 28, 35
water vapor, 7
weakness, 43
weapons, 71
wear, 10, 17, 18, 27, 30, 33, 71
wet, 45
witness, 43
wood, 23

Z

zinc, 7, 8, 9, 55